D0842038

# A NEW WORLD

WHITLEY STRIEBER

WALKER & COLLIER, INC.

Walker & Collier, Inc.
20742 Stone Oak Parkway
Suite 107
San Antonio, Texas, 78258

www.unknowncountry.com

First Walker & Collier printing, first edition, 2019

Library of Congress Cataloging-in-Publication Data
Strieber, Whitley

A new world / by Whitley Strieber
ISBN (Paperback) 978-1-7342028-0-9
(Electronic Book) 978-1-7342028-1-6
(Audio Book) 978-1-7342028-2-3

Cover design by Lisa Amowitz

Printed in the United States of America

First Edition

*This book is dedicated to the children, to whom this world belongs.*

"The great enemy of truth is very often not the lie—deliberate, contrived and dishonest—but the myth—persistent, persuasive and unrealistic. Too often we hold fast to the clichés of our forebears. We subject all facts to a prefabricated set of interpretations. We enjoy the comfort of opinion without the discomfort of thought."

 ---John F. Kennedy,
 Yale University Commencement Address, 1962

"Mythology exists at a level of our social reality over which normal political and intellectual action has no power."

 ---Dr. Jacques Vallee

"We are part of a symbiosis with something that disguises itself as an alien invasion so as not to alarm us."

 ---Terrence McKenna

# ACKNOWLEDGMENTS

I would like to acknowledge the help and support of Lorie Barnes, Josh Boone, Raven Dana, Dr. Jeff Kripal, Anndrea Taylor, Prince Stash Klossowski de Rola, and too many others to name. Their help has been of inestimable value, and I can only hope that I have met their expectations. I would especially like to thank Dr. Kripal, Anndrea Taylor and Stash de Rola for their extensive and patient editorial help, and Josh Boone for reading all seven drafts with such care and insight.

I wish also to honor all the witnesses and researchers who have struggled with the close encounter experience and the effort to understand it for so many years.

I would especially like to thank Jeff Kripal and Rice University for creating the Anne and Whitley Strieber Archive, which preserves thousands of the letters that we received after the publication of *Communion*, and were collected and cataloged by Anne Strieber.

*When Col. Philip Corso asked one of our visitors what was on offer for us if we let them into our lives, the answer was "A new world, if you can take it."*

In Memory

Not a page of this book is absent the influence of my beloved wife, Anne Strieber. Meeting this brilliant human being blessed and defined my life. She brought crystalline insight into the ambiguous and yet real events that we experienced. The byword of her life was "have joy," and it is in that spirit that I have written *A New World*.

# A NOTE FROM THE AUTHOR

Most of what you read in these pages is going to be strange beyond belief. This is because it is about events that are supposed to be impossible, a level of reality that isn't supposed to exist and relationships that are entirely new. Knowing this, I have made every effort to tell my story accurately. I have never left anything out, changed anything or edited anything because it seemed too unbelievable. It bears essentially no relationship to any of the popular narratives about alien contact, even less those of ordinary life. And yet it is, word by word, based on observation and experience.

Unlike many stories that deal with strange experiences, I have attempted whenever possible to add the texture of witness to my narrative. Given what I am asking here, readers deserve to know the degree to which every experience I relate was shared by others.

It is also important to add that the close encounter experience only begins with what we now know as the physical. As you develop your relationship with the visitors, you discover that what we call the physical is only part of a huge tapestry of reality. The vision of those who do not strive to gain from their encounters remains bound to the familiar world, though.

Presently, their physical appearance, although only a small part of what they are, is all most of us know about the visitors. For example, it is my belief that most people operating behind the curtain of secrecy have rarely experienced them except physically, and therefore have a very limited vision of them. But for anybody willing to see and accept the mystery that they actually present, there is so much more. It is this group that has the potential to lead the world to real change.

Engaging with the visitors more deeply is extraordinary and rewarding. It is also completely different from living life as we have come to know it. The laws of reality change. Above all, the very nature of communication changes. The rules are much different and, by our standards, very strange.

I could have edited my story to make it easier to believe —left some things out, changed others to make them seem closer to the familiar than they are. Like the visitors who are part of my life, I hate deception and will have no part in it. To make my story more believable, I would have had to turn it into a lie.

# PREFACE

## This Book Is Contact
## by Jeffrey J. Kripal, PhD

This is a book about a new super natural world in which communication between the visitors and us, between the "dead" and the "living," enters a new level of intensity, where physicality is extended into some other new materialist or superphysical dimension, where time, and so evolution, do not work like we think they work, where astonishing sensing capacities or parapsychological abilities run in genetic lines (otherwise called families), and where the UFO is as much a vehicle of soul as it is a source of metamaterials or the invisible object of new radar returns or classified military attention. There is also a model of embodiment shining between these remarkable lines, constituting a paradoxical vision in which we *use* the physical body as a kind of temporary portal into these physical and temporal dimensions, even as we also remain outside the body and its particular sensory-generated reality. This is a new world not because it is really new (I assure you that all of these themes are very old convictions in other cultural codes), but because it depends on us right *here* and right *now* to take shape and appear.

The deepest message of this little book, then, is an intimate one: that the actualization and appearance, or continued distortion and camouflaging, of this new world *depends on us* and, more particularly, on how we choose to interact with the invisible presences of our cosmic environment. These choices include whether and how we read this very book, which, in the intentions and understanding of the author, is itself an urgent communication from the visitors. The intended implications are clear enough: to the extent that you really and truly interact with and so actualize this book, you really and truly interact with and so actualize the visitors. This book is contact, but this contact depends *on you*.

Allow me to be nerdy for a moment, as this new world and our uncanny reading role in bringing it into focus intersects directly with my daily life and work.

I live and work in an elite academic world, in a school of humanities at a major research university with some of the smartest people on the planet. (I am not sure how I got here, but that is another story.) These remarkable intellectuals see through things, like so many X-ray machines in a doctor's office. They see into the bones, organs, and mostly unconscious structures of whole societies, nation-states, empires, value systems, and religions. Nothing is sacred here.

The fact is that everything human has been invented by humans, so everything can be questioned. The central argument of my own life-work is that there is a secret human potential hidden in such a deep questioning. To understand who we are, we must first understand who we are not. We must understand that the Human is Two. Yes, we are the hairless, dying primate with this or that social identity, but we are also something Other and More that we cannot name with any words or measure with any instrument.

Of course, looking at ourselves in this way can be frightening. This is because everything we think of as "me" or "us" ends up being called into question, seen through, and, in the end, simply set aside. We are left with a mystery, and that mystery is us. But there is also real hope, since we are Other or More than any of the identities we have invented for ourselves.

This is why I so appreciate the ferocious insights of my colleagues. I so admire their unflinching search for deeper and deeper structures of truth and, by implication, for justice (since every structure they uncover and see through privileges and includes some people and marginalizes and excludes other people). I have called this broad-based intellectual sensibility of seeing through societies, nation-states, and religions "prophetic," not in the sense of predicting the future, but in the sense of giving witness to difficult truths that a particular society, authority, or individual does not want to hear. These truths might be about gender or sexuality, about race, about class, about colonialism or empire, about power, about God or the gods, well, about pretty much anything humans think, make, do, or identify with.

But none of this makes academics infallible or all-seeing, and social justice and the endless sufferings of different identities are by no means the only kinds of truth to seek and to know. It turns out that there are sacred cows in the academy, too (and even the Hindu cow, by the way, was once not so sacred in India, not at least as it is today—it was once raised, herded, prized, and, yes, eaten). Our present-day sacred cows, which we will no doubt someday "eat," involve an unquestioning commitment to naturalism, materialism, and scientism, that is, the largely unconscious assumptions that what we think of today as "natural" is all there is; that everything is finally "physical" in the ways that physics understands matter (not very well, it turns out); and that the objectifying methods of science are the only way to know reality, with the often unstated assumption that anything science does not know or cannot know with these same methods must not be real.

Believe it or not, virtually every aspect of modern intellectual life is committed to this triple set of assumptions around naturalism, materialism, and scientism. Take the study of religion, my own area of research and writing. I have spent the last four decades studying religion. For the first three of those decades, I was schooled in a way of thinking that argued, with some very good reasons, that every religious prodigy—every shaman, mystic, seer, saint, medium, or spiritual teacher--is just a biomedical body locked down in space-time, that *everything* he or she claims to know came through a material text, a social institution, or a social interaction. In other words, I was taught to believe that everything human is finally and completely "historical" and "social," which is to say: conditioned, relative, local, and material. Nothing super here.

This is often the case, of course. Hence the prophetic function of the humanities: *all* claims that a religion or nation-state makes to include some and exclude others can be deconstructed, taken apart, shown to be false in any universal or absolute way. That's because they are.

But that is not all there is.

I once thought that all talk of sky-gods inserting special "crystals" or sacred "stones" in the bodies of shamans to signal the calling and authorize the teachings of a new spiritual prodigy was nothing but the stuff of myth and folklore. Then I encountered Whitley Strieber and the implant in his left ear, which, as he explains in his books, was central to his own calling and now inspires him to write his books, sometimes with very specific information. Turns out that the mythologies and folklore were true.

I once thought that scriptural scenes like that described in the first few chapters of Ezekiel could be explained as literary inventions, that every detail of them could be traced back to some other text, social institution, or previous belief. Then I encountered the modern abduction literature, many scenes of which feature apparent craft and seeming technology that look, well, pretty much exactly like Ezekiel's famous "chariot" vision, which is obviously no chariot at all. Turns out that the Bible, or any other religious scripture for that matter, is sometimes describing actual human experiences and not just making stuff up.

I once thought all that talk of "ghosts" and "spirits," of invisible spectral beings, was some kind of psychological projection, social construction, or mistaken dream. Then I met Whitley Strieber, who interacts with invisible "visitors" on an almost nightly basis, most recently to write this book. Turns out that ghosts and spirits are real, "real" in ways that we still cannot grasp or define with our prescribed evolved senses and adaptive cognitive capacities.

I have interacted with Whitley for about a decade now, often on a daily basis. I believe that I have also interacted with his departed wife, Anne, through a most remarkable drawing channeled by a Canadian medium (more on that some day). Perhaps most dramatically, I have slept in the same room with Whitley at a symposium (a risky enterprise, I assure you) and experienced my own psyche "split" in two in his sleeping presence. Some other part of me, completely separate from the conscious self, watched something astonishing take place in that dark room. As this all transpired, this Jeff-self heard very distinctly in his mind the following words uttered by some other part of me: "Oh . . . my . . . *God*!" The tone was one of ontological shock, as John Mack once put it so well: not quite fear, but something that included the intellectual emotions of amazement, astonishment, and a kind of pure or total cognitive dissonance. Indeed, I was seeing something *so* shocking and *so* dissonant that I could not see it. I literally could not process it as myself. And so I split myself in two, so as to process it and not process it, so as to see it and not see it.

Most recently, just a few days ago actually, Whitley shared with me a recording of a similar night scene. The audio recording features the other Whitley (he is Two, too) speaking lovingly to a visitor in the room whom he calls by name: "Teach me, Mature." The recording ends with a female presence sighing deeply as she sensually and intellectually interacts with Whitley. As I listened to the audio, I was reminded of the feeling tone of that night that I split in two in our shared room. Like Whitley, I suspect that this female visitor was in some way "Anne," whatever that social convention we call a name means to that being now.

Clearly, I am not the same person (or persons) I was in graduate school, or even a decade ago. I "have" a social name and a physical form, as Anne once did, but I don't really believe either. And I live toward what Whitley calls here "a new world." Alas, I cannot say that I have entered it, or that I have shaken off identification with my social ego or bionome, but I can say that I have seen and heard entirely too much to believe the old world that I have tried my best to leave behind and in which I no longer believe. Words like "religion," "myth," "folklore," "shaman," "mystic," "possession," "god," "demon," "spirit," "animism," even "body," "mind," "individual," "history," "time," "identity," and "human being" all mean entirely different things to me today than they did a decade ago. That old world is gone, even if this new world has not yet taken shape.

"What?" some of my colleagues might say, "You expect us to believe that invisible beings interact with humans, that the human is also superhuman, and that *this* is what the history of religions really points toward, some kind of post-contact social formation?"

"Yeah, that's pretty much it."

I would assure my colleagues that we do not know what or who these invisible presences are; that I suspect they are in fact us (whatever "us" really means); that I understand the supernatural as super natural; and that I do not believe any of the traditional religious, sci-fi, or military mythologies that get wrapped around these mind-blowing moments of real contact and spiritual transmission. Whatever this new world is, I strongly suspect that knowledge practices like "religion," "science," and "technology" are fundamentally inadequate, and that we will finally have to *be* this new world to know it. All the beliefs, science, and weapons of the world will not get us one inch closer to this new world. Indeed, they will only take us further and further away.

But I doubt that any of these qualifications would help these particular colleagues very much. They are still living in the old world. I choose to live toward and for the new one.

Which world do *you* choose?

Jeffrey J. Kripal
J. Newton Rayzor Professor of Religion
Rice University
Houston, Texas

## 1

## THE MYSTERY BEGINS

I am the author of many books, both novels and nonfiction, most notably the novels *The Hunger*, *Warday* and *Superstorm*, which was the basis of the movie *The Day After Tomorrow*. But my best-known work is a nonfiction book called *Communion*, about a close encounter of the third kind that befell me on December 26, 1985. Since then, my relationship with the strange people I encountered on that night, whom I have come to call "the visitors," has continued. Starting in the fall of 2015, contact with them has exploded in richness and wonder.

In part this is because I have gradually learned something of how they can communicate with us—which is so different from anything one might expect that it took me many years just to understand that it was happening. My hope is that by describing my experiences with this, I can help others to understand the messages that they may already have received and also learn how to engage in give-and-take. If more of us can gain an idea of how it works, I think that our visitors may abandon their long-held stance of secrecy and become a more open part of our world.

My first encounter with them did not go well. Not at all.

On that snowy December night, I woke up in a little room filled with what appeared to be darting, big-eyed insects and stocky, dark-blue trolls. The next day, a vague memory of the big eyes caused me to imagine that an owl had come into our house, but as there was no entry point, this could not have been the answer. I was injured in the side of my head. My rectum was torn so badly that I bear the scar to this day. To say the least, I was severely shaken up. As the weeks passed and I discounted one ordinary explanation after another, I was finally left with only one alternative: as incredible, as impossible as it seemed, the event had in some way been real. The creatures I had seen were actual, living beings of some kind.

I agonized over telling my wife. What in the world would Anne think? Things were rocky between us because, during the six weeks or so after the event that it had taken me to understand that I wasn't going insane, I'd tried to drive her off. If I ended up in a psychiatric institution, I knew her well enough to know that she would never abandon me, and what would she and our son do for money? They needed and deserved to have a healthy husband and father, not be saddled with a person so deeply psychotic that he was completely misperceiving reality.

I took neurological tests for brain abnormalities and temporal lobe epilepsy, which can produce hallucinations. All were normal. The epilepsy test even showed that not only was I not prone to hallucinations, I had an unusually stable brain. I took an extensive battery of psychological tests, which revealed me to be normal but also suffering a high level of stress.

I still had no idea what to say to Anne. What could I say, that I'd been taken aboard a flying saucer by little men? Because I was now pretty convinced that this was what had happened. Finally, I told an old friend, photographer and documentary filmmaker Timothy Greenfield-Sanders. Incredibly, he responded that his wife's parents, who lived down the road from us in upstate New York, had seen some strange creatures similar to the ones I had described to him. They had observed them from a window, moving about in their back garden.

This put things in a new light. I could now say that there were other witnesses. Obviously, Anne was going to ask that question. Timothy advised me to go ahead and tell her. He was sure she'd find it all as fascinating as he did.

One evening after dinner, when our little boy was safely asleep, I asked her to sit down for a talk. She said later that she was afraid that I was going to tell her I wanted a divorce. But I was past trying to drive her away from the crazy man. My main concern now was whether or not *she* would be the one to want a divorce.

I said the words I had been so afraid to utter: "Honey, I think I was taken aboard a flying saucer by little men."

She stared at me. Her mouth dropped open. Then a twinkle of what I can only describe as her sparkling wit flashed in her eyes. She blurted out, "Oh, thank God! I was afraid you were going crazy!"

There was a silence. Her eyes searched me for some sign that I was joking. I looked gravely back at her. Then we both burst into laughter anyway. We threw ourselves into one another's arms. She said, "I wanted an interesting life, but I had no idea what I was getting into when I met you."

In that moment, our marriage entered a new chapter. On that evening in our little apartment in New York, we began what became a journey of discovery that completely revised our understanding of world, life and reality and continues to do so every single day.

From the beginning, she seemed to know things that she perhaps could not put into words, about how this incredible experience belongs first to us—to human beings—not to the mysterious figures in the night who trigger it.

At first, I was appalled. I resolved to sell the little cabin where it had happened and never spend another night outside of the depths of a city. But I just couldn't get over the idea that they were real. *Real.*

Anne didn't want to sell up. She wanted to go back and see what might happen.

I also found that my curiosity was stronger than my fear, and in April of 1986, we resumed our weekends at the cabin. I started going out into the woods at night to serve notice to whatever they were that I wanted to meet them again—hopefully without a repeat of the violence that had taken place previously and the injury that my flailing panic had caused me.

Given what I had been through, this might seem pretty foolhardy, and it certainly was that. But I was simply too curious. We both were. The fact that the event had happened at all had been completely remarkable. Although I had been left with injuries, I had also had my mind opened in the most profound way I could imagine. Whatever they were, they were not human by any definition I knew. And yet they were here.

Anne agreed that I should go out and see what happened, but she was not a believer by nature. She was a questioner, and I followed her lead in this. She would often say, "The human species is too young to have beliefs. What we need are good questions."

We humans have a nasty habit of deciding to believe that things we don't actually understand are explained in some way that we make up. And the next thing we know, we're killing each other over these imaginings.

So this is not going to be a book of advocacy. I am not asserting beliefs. Instead, it is going to be a description of

4

events, as I have witnessed them, and an inquiry into ways of communicating with a richly alive, enigmatic and absolutely remarkable presence that seems to have been with us, at least in part, throughout our recorded history, and probably a lot longer than that. But it is also true that it seems to have entirely changed its approach to us in the years since World War II ended. It is now an enormous part of our world. It seeks to become even more central to human experience. We're dying here, and it doesn't want that. It wants to communicate with us, but it is deeply, profoundly different from us, and so far communication has been essentially impossible. Learning to do it effectively is what my life has been about, and what I hope to convey here.

The visitors have been very clear to me: Unless we can communicate with them in a rational, practical and effective way, they cannot help us.

I wish I could provide a simple how-to, a neat list of dos and don'ts.

I can't do it, and nobody can. The gap between us is simply too great for a simple list to work.

Our visitors stand ready to help us face the jeopardy we are in, and even aid us in solving the problems that we are facing. The degree of their involvement depends on the degree to which we are able to face them and understand what they have to offer us. For reasons that are going to become clear over the course of my story, this is not going to be easy. Far from it, making sense of the relationship will be the greatest intellectual, emotional and spiritual challenge that mankind has ever faced. If we are able to succeed, though, we are going to experience a vast increase in the range of human understanding—truly, we are going to enter a new world.

Right now, things are predictably chaotic. Different religious groups have long ago integrated the phenomenon into their belief systems. Christians think of the visitors as demons, sometimes as angels. To Muslims, they are *djinn*. Other religions call them many other things. Outside of the Christian community in the west, the fairy faith of Northern Europe has evolved into the UFO and alien folklore and has spread around the world. But that is only part of what has happened. What used to be a minor folklore is now a vast, living experience, by far the most complex cultural and social influence, and personal challenge for the close encounter witnesses, that humanity has ever faced. Millions of individual lives have been touched by it. Governmental response has mostly been concealed behind a veil of secrecy that hides disquiet, confusion and fear.

In point of fact, the entire human species has been thrown off balance by whatever this really is. Despite over seventy years of effort, it has not just remained a mystery, it has become steadily more mysterious—and at the same time, more and more provocative.

Somebody is here, all right. At this point, denial of that fact is an emotional response not a rational one.

It is not a simple matter of aliens having arrived. This sublime, challenging, sinister and yet oddly welcoming presence is far more complex than that. It is so varied, so contradictory, and yet so pervasive on so many different levels that for us even to achieve a useful description of it is going to be a tremendous intellectual challenge.

Behind the scenes, a small group of scientists, who have access to certain materials and biological remains, has made significant progress in areas as diverse as metallurgy and communications. But most scientists and intellectuals, left out of this exclusive group for reasons that will become clear over the course of this book, remain in a state of enforced ignorance that emerges into the broader culture as a mix of denial and indifference. Meanwhile, people who have been touched—often, as I was, fiercely—by the presence are left to fend for themselves when it comes to understanding what has happened to them.

The result of this is that the rich potential of contact is being buried beneath a great mound of confused theories that amount to little more than an elaborate extension of the folklore of unknown presences that has been with us from the beginning of our history.

Make no mistake, though: it is not up to the people concealing their knowledge behind the wall of classification to disclose the secrets. From the beginning, the visitors themselves have been in control of the secrecy, and it is they who will be responsible for revealing themselves—or, to be more accurate, integrating themselves into our lives more openly than they already have.

They are extremely subtle, very thoughtful and very careful. The ones I have come to know want contact to work, but if they were to step out into the open right now, that is very definitely not what would happen. Instead, we would try to integrate them into our worldview through the medium of assumptions about aliens that simply are not adequate to the task and do not hold up under scrutiny.

They are also determined, hard and can be, frankly, terrifying. They are powerful beyond conception and are not going to be defeated by military opposition. That whole approach is meaningless. This isn't a war at all, it is a process of contact that is intended to lead to the deep inner sharing that is communion. There are things that they want from us and things that they can give us in return. But it is a trade of a kind we have never engaged in before and it is going to take a leap of understanding for us to make a success of it.

It is not, however, something that we don't do. We have been doing it, I would think, through all the time we have been. We have been passive to it, however. The result of this is that both sides gain less from it than they could. They do not share as fully as they might. We do not realize what we actually are, and so devalue ourselves. Only when we become aware of the sharing can there be communion between us. As we are now, we're passive to it. We need to engage with it and so with them, thus becoming participants in an enormous part of our lives that we have not so far known existed.

This is what it is to increase in consciousness. It means knowing and seeing more of where and what you truly are. As I have said before, the coming of the visitors is what it looks like when evolution comes to a conscious mind.

There are some previous moments of contact in human history that are instructive here. The first two involve how pre-Columbian civilizations responded to the unexpected appearance of the Spaniards, who arrived in the Americas in possession of things like horses, metal armor and gunpowder, none of which were known to the indigenous population. However, they also brought with them a culture debased by superstition, fanaticism and jingoistic brutality. This devastating combination was used to destroy not just the Indian societies in Mexico and South America, but later to subjugate and enslave populations all across the western half of North America. Further waves of Europeans later overwhelmed these ancient Earth-centric cultures over the entire continent. Only recently have any of them begun to show even a few sparks of recovery, and most probably never will.

Later in this book, I will discuss an eloquent message left by the visitors, warning that we could face similar cultural dislocation, although hopefully without the exploitative brutality. This warning has been placed at a site where advanced artifacts were left behind and from which scientists have been recovering valuable materials for years. The reason that this particular site was chosen by the visitors has not been understood so far, but it can be, and its warning should be, heeded carefully at every level of our society as the visitors emerge.

The Aztecs attempted to fit the bizarre apparition of the Spaniards into their existing mythology and decided that Hernando Cortez must be their god Quetzalcóatl, come back to retake the throne of the Aztec empire. Although they were momentarily nonplussed by the fact that he couldn't speak their language of Nahuatl, they decided to ignore this anomaly and more or less allowed a few hundred Spaniards to conquer their extensive empire. The result was that, within 50 years, 90% of the indigenous Mexican population was dead and the surviving 10% were enslaved.

Another instructive case is that of the Inca civilization, which was overtaken by Francisco Pizarro. Essentially, a highly organized nation with an army of 50,000 men was defeated and subjugated by about 500 Spaniards. The critical battle of Cajamarca in 1532 was won by 200 Spaniards, who defeated an Incan field army 6,000 strong. When Pizarro subsequently entered the Inca capital of Cuzco, he was met by what amounted to a confused silence. He and his Spaniards made no sense to the Inca. The result was that this ancient civilization, with probably the most mysterious origins of any on Earth, also disappeared into history, taking with it all that it had learned and achieved, leaving behind only the ghostly remnant that persists to this day among indigenous Peruvians.

In attempting to integrate the Spaniards into their belief system, the Aztecs did what religious and UFO believers are doing now. They looked to their own preexisting beliefs to explain the inexplicable and were rendered helpless as a result. Our scientific and intellectual communities are reacting much as the Inca did, with confusion, denial and, finally, silence.

What happened when technologically advanced European

civilizations expanded across the world should also be carefully considered. Even when they were not brutal and exploitative, the technological superiority of the Europeans again and again made indigenous peoples feel inferior, caused them to abandon their own cultures and beliefs and, all too often, descend into the debased remnants of their once vital societies that we see to this day in too many places in the world.

The visitors are not just technologically more advanced than us. They have a completely different way of seeing reality, and when we first confront it, we are going to see them as being in possession of godlike powers, breathtaking insight and seemingly unlimited scientific knowledge. They will appear this way not because they are more intelligent but rather because of the different way their minds work. They are not more intelligent than we are, not at all. They are more experienced, and their experience is fundamentally different from ours. We must not elevate them above what they are but seek to meet their strengths with our own in the most meaningful and mutually enriching ways that we can find.

Their abilities are going to disempower our scientists and intellectuals, which will be made worse by their inability to address them directly. In this regard, I am hoping that this book will accomplish three things: first, that it will enable people to see that there are ways to communicate with them; second, that we have things that they value greatly and need badly and that a great trade is on offer; third, that they do not want to invade us, enslave us or otherwise destroy our freedom or take from us our own sense of self-worth.

They think highly of us and consider that what we have to offer is valuable. Otherwise, they would not have any interest in us.

Anne named our book *Communion* because contact is about deep sharing not exploitation. She understood this very clearly from the beginning. As the great European empires of the 19th and 20th centuries found out to their cost, colonization and exploitation is a costly and unrewarding business. If this was a direction the visitors intended to take, they would have done so long ago.

By the early nineties, even though I had been more or less driven from the public stage, our relationship with the visitors did not stop. It never stopped. In fact, it has grown from the violent, confused mess that it was originally to the richly rewarding companionship that it has become.

It has been a long, hard road, though. On the evening of August 11, 2015, I lost Anne, whose brilliance and insight had guided us. Lost her, but only in a way, for when you are as deeply involved with the visitors as we are, death does not bring with it the same finality that it normally does in human experience. As I have said, they don't function in reality in the same way that we do, and when you get close to them, neither do you. Anne and I wrote a book about this called *The Afterlife Revolution*, which I will refer to frequently in these pages. We wrote it—together—after she had died. Because of the state we are now in, which I believe to be new to human experience, I wear both of our wedding rings. This is to symbolize that fact that we are still together, only now possess only one outpost in the physical world, which is this old body of mine. I know that this is deeply heretical to our scientific and intellectual communities. This is because the intensity and complexity of material culture has blinded us to the existence of that subtle and many-faceted enigma we call by the single, starved word "soul." But this mysterious and denied presence is quite real. It is not supernatural in any way, but rather part of nature. It is also where our visitors live, penetrating only occasionally into this physical level. As they come closer to us, it is going to become more and more clear that the reality of the soul is much bigger than that of the physical, but also that it is nothing like we have imagined. It is going to become inescapable that not only is consciousness in us, we are in consciousness.

As our relationship with the visitors deepened, it became inescapable to both Anne and me that the physical world is embedded in a much larger, older and richer reality. However, we cannot yet detect it with any instrument we possess. The result of this is that we deny it altogether, in part because we cannot apply any known method of discovery to it, and in part because we in the west fear that our secular freedoms, acquired at great cost from a religious dictatorship that lasted a thousand years, will be lost to that dictatorship once again if we so much as whisper the possibility that the soul may exist.

It is not the soul of religion, though, but something very different, much more real and, in the end, much more obviously part of nature. Nor is it really separate from the body. A living being is, rather, a continuity. The borders of a human being are not found along the edges of the flesh but rather within us, where there are depths that we have forgotten but can regain and, if we are to survive much longer as a species, must regain.

The visitors stand ready to help us refocus ourselves in this new way and, in so doing, make for ourselves a new world, one that will involve permanent physical survival and conscious extension into the larger reality that has hitherto been addressed only with confused belief and, more recently, denial. Doing that is what this book is about, and why it has the title that it does. Truly, a new world awaits us. All we need to do is respond coherently to them to begin what I am certain can be a journey of incredible value to us and all who follow in generations to come, of which I hope with all my mind, heart and soul there will be many.

## 2

## AN URGENT CALL

Over the thirty years that I have had a relationship with the visitors, I have gotten some idea of what they want from us and have also formulated some thoughts about what they have to give in return.

The most important thing on offer is knowledge, and they are already bringing it, but mostly in very concealed ways. We need more of what they have to share.

One example of a scientist who has gained knowledge from them is the distinguished mathematician Dr. Edward Belbruno, the winner of the 2018 Humboldt Prize in Mathematics. Belbruno's books include *Fly Me to the Moon* and *Capture Dynamics and Chaotic Motions in Celestial Mechanics*. He is a consultant with NASA and also a prominent artist.

I met Dr. Belbruno in 2009 and interviewed him on my podcast Dreamland. At the time, he had never made a public statement, so he appeared anonymously. This distinguished man has since gone public with his story.

On October 2, 1991, Dr. Belbruno was driving along an isolated road in Wyoming when he found it blocked by a large object. A short time later, it rose into the air and was gone. He feels that his space trajectory research was affected on that night, and the calculations that came into his mind have been extremely useful in his work.

Unfortunately, I cannot name some other scientists I know who have gained knowledge from the visitors in similar ways, largely because their work is classified, an issue that must be considered now with the utmost seriousness. In view of the urgent need to accelerate scientific, technological and cultural progress, there needs to be an internal review at the highest levels to determine what, if anything, should remain classified at this time. Contrary to many conspiracy theories, Executive Order 13526 establishes three levels of classification, the highest of which is Top Secret. "Top Secret shall be applied to information, the unauthorized disclosure of which reasonably could be expected to cause 'exceptionally grave damage' to the national security that the original classification authority is able to identify or describe."

The first National Security Act, passed in 1947, mandated that conditions that might be threatening be classified until their status could be determined. It was on this basis that the first UFO debris, which was found near the Roswell Army Air Force base in July of 1947, was classified. Initially, this had nothing to do with aliens. The 509th Bomb Wing, stationed at the Roswell base, was the only atomic bomber wing in the world. It was also the only thing preventing Stalin from invading western Europe, which would have resulted in an allied defeat within six weeks. So the appearance of some sort of unknown flying vehicle near the base was deemed a threat, and it was classified as soon as Army Air Force authorities became aware of it.

General Arthur Exon, who was a friend of my uncle Col. Edward Strieber, told me in 1988, "Everyone from Truman on down knew that what we had found was not of this world within twenty-four hours of our finding it." Both were at the

Air Materiel Command at Wright Field in Dayton, Ohio, where the debris and biological remains were brought for study. My uncle said that the debris he had observed had properties identical to those described by Col. Jesse Marcel in a video he made in his later years stating that some of the metal they found "wasn't any thicker than the foil out of a pack of cigarettes, but you couldn't bend it, even a sledge-hammer would bounce off of it." My uncle told me that it was subjected to bullet tests, and bullets could not penetrate it.

Both my uncle and General Exon told me that they had been lawfully requested to give me the information that they did. They were not in violation of the law when they spoke to me. I do not know who might have made this request of them, nor would they tell me.

Since 1947, the amount of information held secret about this and many other events related to UFOs has grown and grown. The reason is, basically, that the law needs to be changed. Right now, unknowns such as the Roswell materials and biological remains can be held secret until they are not deemed a threat. The language should be that such things may only be held secret *if* they can be deemed a specific threat. While this change will not lead to the release of everything held secret regarding this subject, it will require the release of information about materials such as those found at Roswell, as well as biological remains.

The military keeps the broader matter secret for three reasons. The first is that the visitors cannot be controlled in any way. We are militarily helpless. During the period 1965–2000, when abductions were frequent events, the fact that they could neither be understood nor stopped caused an increase in the intensity of the secrecy. The fading of the abduction phenomenon has not changed this. The second is that many processes we have already learned from the materials we have obtained can potentially be weaponized, and in fact, there is a race on between Russia, China and the United States to do just this. Central to this hidden competition is discovering the secret of gravity, the mastery of which is displayed by the UFOs. Therefore, research into their motive power is held deeply secret. However, it is also true that none of the players have won this game, and perhaps it is time the broader scientific culture gets its chance to address the mystery. A good start would be for more materials and UFO footage to be released and for the National Academy of Sciences to change its opposition to granting in this field.

The third is that the visitors themselves compel a degree of secrecy because, as matters stand right now, we do not have the intellectual tools we need to communicate coherently with them. They do not do this via some sort of hidden conclave. They do it by cultural and social manipulation, and they are really very good at this.

What happened at Roswell is a good example. The crash, which, like the others, was probably more a donation than an accident, took place within thirty miles of the most secret military facility the United States possessed, at a time when it was the primary key to the preservation of freedom in much of the world. Of course the Army Air Force jumped to keep it secret. Then, when they realized what they had, complete confusion descended. Over the next few years, as the military came to understand that it was helpless before this apparent threat, the whole matter remained classified. Meanwhile, efforts to understand and make use of the secrets of the remarkable debris went into overdrive, as did our weapons research.

In other words, we did exactly what the visitors wanted us to do, which was to keep their presence secret while we caught up to them technologically. They are not interested in supplicants, let alone slaves. They want us to be independent, self-sufficient companions. Otherwise, we are more or less useless to them. Over the course of this text, I will explain as clearly as I can why I am sure that this is so. To conclude now, though, I can only say that, with the capabilities that they display, they could have invaded us and subjected us to their control years ago. I doubt that the subjugation of the whole planet would have taken them more than a week. I can easily believe that it could have happened in a matter of seconds. But it didn't happen, and because it hasn't, I think that it's safe to say that they have another motive.

I don't think that they would have spent thirty years teaching me or anybody else the rudiments of communicating with them unless they were here to do just that. I think that their motives are to contact us, communicate, and finally enter into the deep sharing of communion.

Communicating with them is not like communicating with each other. Not at all. But I think that it is essential that we turn toward them and try to engage in a meaningful way. We have never so far done this, not in all the thousands of years of history that, to one degree or another, they have been present here.

Because of the way nature designed us, we've overpopulated Earth, with the result that the environment is breaking down. A worldwide failure of political leadership and social will has caused us to ignore the problem for too long. It is my belief that our visitors do not want us to sink into the chaos that now threatens, let alone go extinct, which also seems like a possibility. This is not entirely altruistic. They also want something from us. If my life with them is a true example of what relationship in general will be like, then I know what it is. There will be great challenges for each one of us with whom they engage, and extraordinary rewards for all mankind if our relationship with them flourishes.

In a sense, my 1985 experience was initiatory, introducing me, as it did, to an entirely new reality that had heretofore been hidden from me. It overturned my understanding of the world entirely.

The relationship continued, and I wrote two more books about it, *Transformation* and *Breakthrough*.

In September, just a month after my wife's passing, the visitors burst dramatically back into my life. I can't prove it, but I sense very strongly that this was a result of something she was doing—after death.

Sometime in the early nineties, she said what I believe to be the single most important thing about the experience that has ever been said. One afternoon she walked out of her office after reading letters for two or three hours and stated rather quizzically, "I think this has something to do with what we call death." She said this because we were receiving a steady stream of letters involving the dead appearing along with the supposed aliens. On the occasions at our cabin when the visitors came, the human dead were generally also with them.

Within two hours of Anne's passing, she began making it clear that she still existed and was aware of this world. She did it not by contacting me, but by asking friends to call me. The first of these, Belle Fuller, had no idea that Anne was dead when, at about 9:30 in the evening, she heard Anne's voice in her ear say, "Tell Whitley I'm all right." When Belle's call came, I was sitting in the living room absolutely bereft, helplessly asking her for some sign that she still existed.

She continued to make contact with me through other friends and acquaintances, never directly. By the end of the month, it was undeniable to me that she not only still existed, but that she stood ready to continue our relationship.

I think these events foretell a fundamental change in the experience of being human. Right now, we cannot reliably engage with our dead. Our whole religious journey is, at its core, an effort to ensure that the death of the body does not mean annihilation. One of the most basic changes that relationship with the visitors suggests is that the barrier between us and our own dead is going to fall. Empirical evidence will emerge that enables us to escape the trap of believing that physical life is the only life. On the other side of the wall of death, there lies a new freedom and a new life, and along with them opportunities for enrichment that are presently almost beyond imagination.

My observation of the visitors suggests to me that they exist primarily in a nonphysical state. Not that they don't have physical bodies, they do, but they are not bound to them in the same way we are. This is why something like the Roswell donation could include bodies. They were not people but containers that people used.

It isn't that the soul level is not also material, though. It is, but in a way we cannot yet detect and measure. Like everything, it is part of reality, by which I mean everything that is material or energetic. Secular culture is correct to believe that there is no supernatural. However, it is mistaken in dropping the parts of the natural world that it cannot yet detect into that basket simply because they have so far evaded measurement.

If you find the idea that we continue after death hard to believe—or perhaps to face—I urge you to explore the enormous literature that exists on this subject. There is an energy that we have not yet detected, upon which relationship with everything in this other reality, including our visitors and our own dead, depends.

As has been true throughout the history of science, the existence of any aspect of reality that is not currently measurable is always hotly denied. There are two reasons for this. The first is that the brain is constructed to see what it can use, not what may be behind it, which could be a very different matter. So we have to feel our way toward the truth using instruments of detection. Without the right instrument, we can't detect something that might be quite real, and all too often we fall into the trap of denying that it is there, even when there is evidence to call this assumption into question. The second is that science has had tremendous success with the strategy it is using now, which has reinforced our tendency to reject what we cannot detect as nonexistent. Why look for things that probably aren't there when we are learning so much by studying the things that are?

From observing the way the visitors communicate with me, I have gained the impression that they do not see the world as we do, and in the most fundamental possible way. We use what I call an output strategy. Our senses provide us with a richly detailed view of the world around us. That is to say, we see what's on the surface, but the surface is not where reality begins. We must use instruments to detect what's beneath it. We see an apple but must use an electron microscope to observe its molecular structure, and even more sensitive instruments to detect the atoms that make up those molecules. We can extrapolate the math that organizes them, but we cannot see it. By contrast, I think that there is evidence that the visitors see the principles first, the apple only later. They use an input strategy in order to organize reality in their minds. In other words, they think the same way a machine does. That they may, therefore, be machines doesn't seem to me impossible, either, nor does the idea that a machine might be conscious.

I believe that we are well on our way to creating conscious machines ourselves. I see consciousness as what happens when a mind regards itself, as every mind must. In other words, there is a sort of reflection of us within ourselves, and that reflection is what makes us feel a sense of self. If true, then we can certainly develop a machine that does that, and I wouldn't be at all surprised if others in this universe have not already done so.

Understand, I don't equate soul with consciousness. I have not only been out of my body, I have been seen by others and in one case communicated with the person observing me, so I cannot think that I am entirely confined to my body. When in that state, though, my consciousness isn't the same as it is now. It is less informed about who I am, is the best way I can think of to express the difference I have observed.

I can never forget the moment in February of 1986 when I turned away from a group of visitors who were waiting for me in a clearing near our house. The moment my hand touched the doorknob to go back in the house, there came from above the woods three cries, which I have described as the most perfect and yet most emotionally complex sounds I have ever heard. If a machine had emotions, this is what it would sound like.

On other occasions as well, I have noticed a strange sense of perfection attaching to sounds and movements made by the visitors. It is really uncanny, and every time I see it, the idea that I am dealing with a conscious machine comes to mind.

One of them once made a very telling comment to a close encounter witness. It told her, "We rearrange atoms."

This is the holy grail of technology. If we could do that, we could make anything into anything. We could create completely novel materials. We could, in effect, do anything. I think that they may be able to do this naturally, because of the way their mental processes work. If they are seeing the world, as it were, from within, it's not too hard to believe that they might also be able to change it from within as easily as we rearrange flowers in a vase.

Another suggestion that this may be true comes from that much maligned phenomenon, the crop circles. Of course some of them are manmade. There's no secret about that. But many are not—which, because of how unlikely an unknown origin seems, is a matter for debate. Some of the earlier formations, such as the Julia Set laid down beside Stonehenge one afternoon in 1996, followed shortly after by the Triple Julia at Windmill Hill the same year, were intricately devised fractals. Many formations reflect geometric principles and mathematical formulae, but these early ones, astonishingly intricate, are hard to disprove as anomalies. Like almost the entire body of such formations, they illustrate math and geometry. Only a very few are images.

Is this because they are being made by somebody who thinks first in math and, therefore, is attempting communication with it, but in ways that we can see with our output-related brains and, hopefully, interpret? It seems possible to me. Perhaps it also explains why the first scientist not hidden behind the wall of secrecy who has received information from them is one of the world's leading mathematicians and received his information in the form of a mathematical formula.

One approach is not better or worse than the other, but they are going to result in radically different ways of understanding the world. If I am right, then this difference is fundamental to the communications gap that exists between us and is likely to be the primary reason that they are not more present in our lives. They cannot communicate coherently with us until we—and by this, I mean both sides—have a clear idea of what the differences are between the ways we see and understand the world.

The reason we use an output strategy is that our dense, complex bodies need to see the world in a practical, useful way. If we utilized an input strategy, we might have a more accurate view of reality, but we are not going to be able to find food very easily at all. Perhaps our visitors, less dense, don't have many physical needs, maybe none. They thus have had no reason to evolve an output-first strategy. Based on my observation of them, of how they react to me and how they reacted to so many people who wrote us letters about contact with them, I think it's possible that that they really do see the underlying truth first, then—and only if necessary for some functional reason—the outcome. This would mean that they would see first the forces and math that lies behind the apple, and only if necessary the actual form.

I have no way of proving this, but when I see the differences between how I and they react to the world around us, I think that it could be the fundamental reason that we are so different. To attempt to bridge the gap, they also transmit information through images and by making reference to natural processes, using them as examples of what they mean to convey. In other words, they communicate in a very ancient way, which is similar to ancient Egyptian hieroglyphics. It is as if they must find things in the world around them that fit the concepts they are trying to communicate, and I don't think that they would need to do that unless they were translating a vision that is founded in the underlying structures of reality into one that has meaning at the level of outcome.

If this is true, then it is this fundamental difference in the way our minds work that is generating the massive communications gap that we actually see. As will be seen later, they have come up with some clever strategies that do allow reliable and even rich communications, but they don't involve sitting down across a table and conversing as we might among ourselves.

As this book is about beings and things that cannot now be detected and measured, it is about rejected knowledge. That doesn't mean that it isn't true, though. Voltaire dismissed fossils as fish bones tossed aside by travelers. The possibility of heavier than air travel was definitively denied by Lord Kelvin in 1895. Similarly now, the existence of the unknown that defines my own life and the lives of many others is denied.

It's too easy to attribute this simply to the nature of intellectual culture. It goes deeper than that, all the way back to the issue of our use of output strategies.

In my experience and that of so many people, the visitors are real. The dead are conscious entities. Communicating with both is neither a mystery nor a miracle. But because there is nothing here to measure, at least not yet, it's impossible to prove these things using established scientific method. Therefore, they cannot be entered into the canon of events that are considered "real." My intent here isn't to offer that proof. It will come in its own time and in its own way. What I can do is describe how I have learned to use it and how this has enabled me to open access for myself to aspects of the greater world of which we are a part.

Looking back on what I was before I found the visitors in my life in comparison with what I am now, it is hard to recognize the plodding, frightened, wary man of those days. Their coming devastated me but also freed me. Had I not decided to challenge the fear that the 1985 incident left me with, I would never have been able to move on. I had been violated, and I was angry. But walking out of Dr. Donald Klein's office after the session with him that brought it all together, I thought to myself that I wanted to continue the adventure. I just wanted to.

When I published *Communion*, it was easy for the close encounter experience to be scorned. While it can still be dismissed, it is now much harder to do what was done then and claim that the witnesses are simply delusional. The letters that Anne collected tell a story not of mental illness but of normal people having a tremendous variety of unusual experiences, which all center around one basic event: the approach and presence of small people with large, staring eyes and outsized heads, often accompanied by those dark blue trolls. As often as not, as happened in my case, dead people who appear to be alive are present as well.

The descriptions people have left of what happens to them next forms one of the most complex narratives of human experience ever recorded.

But is there any hard evidence? Any at all?

Aside from the many cases of multiple-witness experience, including some stellar ones at the cabin where I had my experiences, I cannot point to anything so telling as a video of aliens carting somebody off. On a personal basis, I have tried for years and in many different ways to obtain video or photographic evidence. I have studied many videos and had some particularly promising ones professionally analyzed. There are three that are compelling, two of which are almost certainly authentic. Neither of them portrays the sort of entities that are described in the literature, and that so many of us have seen. Instead, they show extremely strange looking stick-figures walking carefully along. The videos can be found by going to Unknowncountry.com, opening the search engine, going to the Out There section and searching on the term "stick figure."

While genuine footage of otherworldly entities would appear to be quite rare, this isn't to say that there exists no hard evidence of anomalous objects or materials that could be related to the close encounter narrative in some way. There is a great deal of UFO video, including video of similar objects taken at different places and times around the world. Moreover, releases of US Navy video over 2017 and 2018 has affirmed that UFOs are a genuine unknown phenomenon. In September of 2019, the Navy stated that these are indeed unknown objects. But are they spacecraft belonging to aliens?

Possibly. They may also be intrusions from another reality, which is an important possibility to explore. Or they may be something so unknown to us that we simply cannot identify it in any meaningful way at present. I don't see much evidence that they belong to an earthly foreign power. The reason is that there is so much good video going back years. If any country had devices of such incredible power, our world would not be as it is now.

In support of the idea that another reality might be involved—a parallel universe—there exist in the public record some high level analyses of metal from a UFO that has proved to have isotopic ratios that not only are not of Earth, but that cannot be generated in this universe at all, not unless there are areas of it operating under such different laws of physics from this one that they might as well be another universe.

What we are encountering might be aliens from this universe, sure. Having laid in their arms and gazed into their very strange faces, I could believe that. However, after all these years of experience with them, while physical aliens may be a component of the experience, I see it as being primarily something that has a different relationship to reality than we do. In order to avoid identifying them with a label that may not be accurate, I have always called them, simply, visitors. Throughout this book I will continue with that convention. I think, though, that there is also a greater presence behind them, or perhaps something entirely different, that has played an even larger role in our development than they have. This I will call "the presence."

Engagement with the visitors amounts to a new kind of experience. There is a gap between us and them, which, I think, has to do with the difference in the way each side is part of reality. If they have evolved on another planet in this universe, they can certainly be expected to look, act and think differently from us. Add to that capabilities such as the ability to become invisible that seem like magic to us, and the issue of how to understand them becomes very large. Are they aliens with skills such as the ability to become invisible, or are they not from this universe at all and using some natural means or technological skill to penetrate into it?

Those are the sort of questions that we cannot really address without more data, but until the wall of secrecy that surrounds what we do know is breached, at least to some degree, we are not going to be able to address them meaningfully.

On rare but important occasions in our history, the larger presence that seems to be behind all this has made an appearance that has changed the world. All of these instances have had to do with religion and have involved the establishment of the idea that there is a single god. Because they have involved flashes of light, strange fires or shining beings, I call them "incidents of light."

The journey began when the Egyptian Pharaoh Akhenaten came to believe that the sun itself was god and was the only god. This was followed by a number of more direct interventions.

Sometime around 1500 BC, a Persian, Zarathustra Spitama, who was wandering in the countryside seeking enlightenment, came upon a shining figure that called himself Vohu Mana, who proceeded to teach him that god, while a single entity, was divided into two forms, order and chaos. Zarathustra spent the rest of his life trying to get people to worship order. Somewhat later, Moses encountered the burning bush, which directed him to lead the people of Israel into the land of Canaan. Then Paul the Apostle, shortly after the death of Jesus, encountered a glowing orb on the road to Damascus and became the first gentile advocate for Christianity. Later still, Mohammed found himself being taught by a glowing angel in a cave, and Islam was born.

While I wouldn't advocate that any of these incidents of light happened as described in the ancient sources, they represent the importance of light to the human mind and, I believe, reflect some sort of deep intention that seeks to bring to mankind an ever more coherent vision of the sacred, and a more and more useful understanding of morality.

Unless the incident at Fatima in 1917 when the sun danced before 50,000 people qualifies as an incident of light, for the most part we see individual creatures, generally small and surpassingly strange. Our visitors can and do take us into their own eerie realm and have been doing this occasionally for eons, as the many stories of such abductions in the literature of the fairy-faith will attest. Something changed dramatically in the mid-20th century, though, as now we have not a few stories of enigmatic abductions but thousands and thousands of them being recorded by people all over the world. This is not a continuation of the fairy-faith. It is similar, but the scale is much larger. It's huge.

This is the reality we are now facing. It leads to the question that forms the heart of this book: How are we to understand this? What is contact, and what sort of change is it bringing to us and our world? For it is bringing change, most certainly. Great change, even staggering change.

# 3

# WHY CERTAIN PEOPLE?

I s it that only certain people notice the presence of the visitors, or that only certain people are approached by them? One thing that became clear to Anne and me when we were getting letters was that contact is a family affair. Often, the experience runs in families. Surprisingly, there were some reports of witnesses being told during experiences that they were members of the visitors' family, or they felt that this was the case.

Sometimes, though, the experience simply comes out of nowhere. An example came from a witness in Australia, who wrote, "In 1976 I was vacuuming my living room floor at about noon. I felt quite ill and thought I was going to vomit, so I sat down on the couch to see if the sick feeling would subside. I then saw that I was not alone; there were three strange little people standing alongside the couch, just looking at me. Two of them were short and fat, about four to four and a half feet tall, with broad faces and enormous black eyes, but with only a hint of where a nose or mouth might have been." She goes on to say that they were wearing brown clothing and seemed to her to be workers. There was another one there who was about five feet tall and "wore a black shroud and had black wispy hair at the back of her head. Her face was very elongated with huge, dark, piercing eyes..." They demanded that she go with them. When she resisted, there followed a sort of mental tug of war, which finally ended when she thought her husband had come home and saved her. But when he really did return, four hours had inexplicably passed. She concludes with a statement that is also a cry from the heart of every close encounter witness I have ever known, "I wonder, where did that time go?"

Nothing ever recovered it for her, no discussion, no regressive hypnosis, nothing. She died some years later with the mystery still unsolved.

A third type of encounter that happens from time to time is one that is intentionally induced by the witness. I first heard of this from Marie "Bootsie" Galbraith, who had prepared a selection of cases that was distributed as *The UFO Briefing Document: The Best Available Evidence,* financed by Laurence Rockefeller and republished in my Dell Books *Hidden Agendas* series. Bootsie told of some successful efforts to "call in" UFOs using laser pointers, but these did not lead to close encounters.

Sometimes they do, and if communication becomes more orderly, I think that this might work more often. However, it comes with a warning. In her book *Extraterrestrial Contact*, Mutual UFO Network Director of Experiencer Research Kathleen Marden reports the case of "Matt," who became curious about UFOs after seeing one on the runway of the small airport he owned. He first signaled his interest by flashing an old navy signal light then by making an X with red rope lights on his runway. This worked to an extent, but he got the strongest reaction by aiming a laser pointer into the sky and clicking it on and off. He was soon seeing craft on his runway, but when he pointed the laser light at them, they would drop slightly then rush at him, flashing multicolored lights. His final experiment with this caused hemorrhages to pour from his nose and ears and ended his willingness to try this approach.

However, the visitors remained interested in him, and he found himself confronting the entities themselves, creatures similar in appearance to what the Australian witness saw. He could not stop the visitations, which became increasingly menacing. He began sleeping with a pistol in his hand. One night, he woke up to see a six-foot-tall entity standing at the foot of his bed. He fired the pistol at it, causing it to disappear in a flash of light, leaving behind a yellow substance spattered on the floor. Unfortunately, he was too distraught to collect it for analysis. There followed a shift into a sinister haunting, elements of which, like some of the events on his runway, were witnessed by other people.

After his family staged an intervention to force him into psychiatric treatment, his mother witnessed, with him, a different version of the world at the end of his runway, a savannah where wooly mammoths were grazing. A year of bizarre events and visitations made him feel as if he were living in a sort of hell. He began to see horrific entities and suffered a profound decline in his health. Just as Kathleen Marden was attempting to organize support from a sympathetic minister, he was hospitalized and died.

The lesson here could not be more clear: If you get curious about this and seek to bring the entities into your life, know that they are not like us and do not have the same connection to reality that we do. Reacting toward them violently without understanding anything about them, even what they are, is foolhardy.

Unlike us, when they are killed, they don't lose contact with the physical world. They can keep acting in it from the nonphysical level. Matt took the body of one of them, but not his access to the physical world. What happened as a result was essentially an act of anger. This individual had lost what they view as a gateway—a body—much more than they see it as a "self." We live by the illusion that we are our bodies. They do not. We assume that the death of the body ends an individual's access to physical life. It doesn't, and if they are attacked and hurt and angered, they are likely to continue to act in the physical world against their attackers, and this is going to look to the attacker like the sort of bizarre haunting that Matt experienced.

This has also happened to thousands of military people over the years, individuals who have acted aggressively toward the visitors. They carry out their aggression under orders that are given by authorities who have little understanding of what they are doing and the consequences of their actions. These military people are being asked to "defend" against them, which is not possible if you have no access to the nonphysical. They are not told that they and their families may experience things similar to what happened to Matt, and certainly not told that their souls will also have to confront those they have attacked after they die, because this is not understood on an official level at all.

Billionaire Robert Bigelow, who has worked with elements of the US government who present a hostile face to the visitors, commented on 60 Minutes on May 28, 2017 that he had experienced a face-to-face confrontation with them. It was not a pleasant experience for him. Similarly, two members of the board of directors of the To The Stars Academy, an organization devoted in substantial part to aggression against the visitors, have had terrifying direct experiences in their lives. One of these led to a family tragedy. The other, fortunately, was only frightening. This organization, which has a rock-and-roll personality as its public facing presence in order to gain easy access to the media, is dedicated to moving the struggle to create weapons that might be effective against the visitors into a more open context in the scientific community, in the hope that some sort of breakthrough can be achieved. Over all the generations of effort that have been expended behind the wall of classification, this has not been a successful effort.

It has failed because defense on a physical level is as meaningless, quite frankly, as it is for a dog to bite the boot of the man who is kicking it. Early on in my adult relationship with the visitors, I had a conversation with one of the blond people who are not of this world and who have also advised our official level. He explained to me that "if you start a war with the grays, they will never let you stop fighting and they will never let you win." This is what is happening on our hidden official level, but it is certainly not what our leaders are told in a briefing that they receive if they ask intelligence and military officials what is being held secret about this subject. They are told stories of what has happened to our military personnel, but not that we ourselves instigated the conflict. Some of these stories are truly horrific. It is then explained that, while we are working on weapons, we have not yet achieved control and the whole thing must be kept secret until we do. It never occurs to anybody given the briefing to ask why it is that the visitors keep matters secret, too.

From where I sit, the official reaction, as inept as it is, is entirely understandable. The visitors are fantastically frightening. There simply isn't any other way to put it. During the summer and fall of 1985, when I knew, but not consciously, that they were there, I also armed myself and might well have done the same thing that Matt did and the military does, had one of them appeared in my gunsights. I would assume that the results would have been similar. Fortunately for me, I was never face to face with them while carrying a weapon, and before I began to understand why they were here and what they want.

Once I realized that they were real, my curiosity overcame my fear, and the guns were no longer a factor. After the initial period of friction, we found common ground in spiritual search. They reacted to my nightly meditations and shared in them. Over time, this has created a fruitful relationship that has enriched my life, and I hope also offered them the reward they seek.

At this point, my relationship with them is as coherent and organized as it would be with anybody I might get to know in the normal course of my life. There are differences, though, chief among them the fact that companionship and communication are so different from what they are among us. They are frightening to me and to some extent always will be. This is because they can control souls, and my soul knows this and in the face of such awesome power, feels the fear of the helpless and the vulnerable in the presence of the aggressive.

I am in my relationship with them to learn and report what I have learned. They are in it to teach. Their aim is very clear: They want the relationship to work, and not just for me. They want it to work, period. They also know how hard it is going to be for us—and for them—to achieve this goal. We are like two brilliant animals of different genius species confronting each other. They are large, we are small. Both are fierce, both are wary.

So what's next? Do we continue circling one another or do we find common ground? Given that we humans are helplessly using up our planet's resources, we are essentially back against the wall. Either we succeed in this or we go where nature and the failure of our environment are liable to take us, which will be into a state where our planet is largely uninhabitable and we no longer have access to physical bodies and all the opportunity that they offer evolving souls.

I believe that my own relationship is in part related to the fact that older members of my family were involved in military operations connected with the visitors. While I don't have any statistical evidence to point to, I do have reason to believe that this can be a factor. Once one member of a family becomes involved with them in some military context, whether confrontational or not, they seem to have a tendency to follow that family line.

My uncle was involved in the Roswell Incident and my father may have had something to do with the presence of Col. Guy Hicks living a block away from us. Col. Hicks was the commanding officer at Goodman Field in Kansas when Capt. Thomas Mantell, stationed there, crashed in his plane in 1947 while pursuing a UFO. This was all debunked, but I think that my father and an FBI agent who lived down the block from us were watching Col. Hicks and some other men on the same street. I cannot say that my father was in the intelligence community, but I have always sensed that he was somehow involved. He was, if anything, even more secretive than my uncle.

Another reason for involvement is childhood trauma. As Dr. Kenneth Ring showed in a study of close encounter witnesses published in his book *The Omega Project*, psychologically stressful experiences in childhood, such as child abuse, also increase the chances of close encounters happening. I think that this is because the abuse shatters the child's expectations, increasing the chance that things that are not supposed to be there, but are, will be seen.

Between August and October of 1952, I was involved in a special education project at Randolph Air Force Base that utilized something called a Skinner Box in enhanced learning experiments. This device, developed by psychologist B.F.

Skinner, was intended to isolate subjects and induce rapid and enhanced learning.

The process was extremely stressful to me, as a result of which I also began having trouble in my regular school, becoming terrified of my teacher for reasons that I could not explain. I began to experience one bout of sickness after another. My pediatrician discovered that I had almost no white blood cells, and my sister and I were removed from the program in October. I was isolated at Brooke General Hospital and given injections of gamma globulin, then kept at home until January of 1953, when I returned to second grade.

My direct memories of this experience are quite confused, and my sister in her lifetime did not remember it at all, but a close friend does remember the Air Force couple who did the recruiting visiting their house. In fact, his parents socialized with them a good bit but refused to allow him to enter the program on the basis that it involved the use of a Skinner Box.

I remember dreadful close confinement, darkness and awful screaming that went on and on. In fact, if my friend didn't recall the recruitment pitch he listened to with his parents, I might think that it had all been a nightmare and that my whole problem was my fear of the nun.

So I had two factors that can lead to abduction: unusual stress in childhood and being part of a family that includes older members who have been involved in some way.

As I described in my book about my childhood, *The Secret School*, it appears to me that the visitors showed up in my life sometime after I was removed from the program at Randolph and remained until I reached puberty, when they withdrew, or I stopped being aware of their presence. By the time I met them in late 1985, I had no memory at all of my childhood encounters.

It is my belief that the depth and extent of my relationship has been enabled not only by my serving notice in those night woods that I was interested in taking matters further but also by something that I have been doing since 1969–1970 called the sensing exercise.

In the fall of 1969, I began working as a Director's Guild Trainee on a movie that was being shot in New York. One of the other production assistants was something new to me, a spiritual seeker. He introduced me to P.D. Ouspensky's book about G.I. Gurdjieff and the Gurdjieff Work called *In Search of the Miraculous*. Anne and I both bought copies and read it with growing interest, and then hungrily. It made the argument that the human attention is mostly drawn automatically to experiences and sensations, but that it can actually be controlled by the individual, who can intentionally split it between what is coming in from the outside and his inner being. This is called the double arrow, looking out and in at the same time. When we experimented with this, we discovered that it did indeed cause a definite expansion, not exactly of consciousness but of what I would call sensitivity. Those early experiments were the first small notes of what has since become a resounding chorus in our lives that crosses the bridge between the worlds and opens the mind to mysterious new levels of reality.

Early on, we found one immediate change that was really very engaging to us. We began to see the world and our place in it in a new way, as both participants and observers. Life became more vivid and poignant. The experience of noticing in this new way transformed ordinary life into an extraordinary art form, and being alive came to appear as a continually unfolding miracle enclosed in the familiar world of our ordinary reality.

We joined the Gurdjieff Foundation's New York City branch. It was in group meetings there that we began doing the sensing exercise regularly. It seeks to split the attention by anchoring part of it in physical sensation while letting part of it take in outer impressions.

Now, fifty years later, I am still doing it, and it is the center of my relationship with the visitors. I am no longer actively attending group meetings in the Foundation, and I don't hold group meetings myself. The Gurdjieff Foundation is the inheritor of the teaching, and I regard work within it as essential to gaining useful understanding and establishing a strong personal practice. I must add, though, that the Foundation doesn't have anything to do with contact or facilitating contact. In fact, the members I've kept up with regard my stories as pretty quixotic. It is a resource designed to awaken us to reality in a new way. The fact that the sensing exercise facilitates contact is entirely coincidental to the Foundation's aim.

On the surface, the sensing exercise is quite simple. It is, however, a uniquely powerful form of meditation, as I now understand. To do it, one sits and places the attention on physical sensation. I follow the method I learned in the Foundation, which involves being attentive to first one foot, then the lower leg, then the whole leg, then working up the other leg, the arms and torso, opening my attention to the sensation of each part of my body until I finally spread it over the whole body.

When Anne was alive, we often did it together. It can also be a powerful experience in a group setting, which the Foundation excels in providing. ·

When our son was born, I stopped doing it for a time. I was too busy and tired, being the father of a baby. But as he grew up, I returned to the practice, doing it nightly at eleven. I now do it at eleven and three, and there is an interesting story behind the addition of the wee-hours session.

Little did we know in the early days that we were not alone—that doing the exercise made me and Anne visible to very different eyes. I have recounted in other books (*Super Natural, The Afterlife Revolution*) how strange things began to happen to us as soon as we came together, long before we had any inkling that such a thing as a close encounter of the third kind was even possible. I will mention here that the first of those events probably took place about six months after we started with the exercise. They involved ghostly events in our apartment in Manhattan and a notable experience of seeing *kobolds* apparently kidnap a man who had gone into a disused storefront where prostitutes sometimes sat. We called the place the "whore store" and used to laugh about it. We were astonished by the sight of what looked to us like blue dwarfs pulling a wildly struggling man behind a curtain. It was terrifying, and we ran home. We didn't call the police because we couldn't figure out what to say. We never passed that storefront again.

Another event from those early days that was important seemed at the time like a dream. This involved being in another world. I was with *kobolds*. We walked along a path under a great arch. Ahead on a ridge, I could see tall trees that looked like cedars of Lebanon. To my left, the land fell off into a vast view of a trackless desert. Perched on the bluff overlooking this desert was a round, dark blue building that was very tumbledown. I was told that it was a university but in bad repair because "the scholars are no good at maintenance." I went toward it at once, eager to matriculate. I wanted to study at a million-year-old university, for sure!

As I approached, I was stopped by peculiar creatures with large black eyes, whom I would later understand as grays, and who I would meet again in December of 1985. I was terribly disappointed when they denied me entry. I have later come to understand that this was actually my matriculation, and that the disappointment I experienced was meant to test me, either spurring me on or causing me to give up.

Creating this type of friction is central to the way the visitors teach, and as a matter of fact, its use is also important in the Gurdjieff Work. Mr. Gurdjieff would create situations that would challenge his students' egos and force them to face themselves, either driving them away or spurring them on. Gurdjieff called it puncturing the "hot air pie" of ego. My hot air pie has been leaped up and down on for many years by rude little men, and I am much the better for it. As I will discuss later when explaining how building a strong soul enables one to render moot the fear of them, the first lesson they ever gave me was about the danger of arrogance and the importance of humility.

I think that I've been in their school ever since that dream. When it happened, the two forms involved were nowhere in the news and not in my life. But there they were, kicking me out of the university that I have been studying in ever since!

Just as maintaining the double arrow enriches one's life experience, doing the sensing exercise enables the sharing of self. While you are doing it, they can enter the silence of your mind and join you in your life experience. For them, this is more than a pleasure. I suspect that, when they are in their normal state, it is an utter delight for them, and I would think that they want to experience it with as many of us as possible. I suspect that Anne intuited this early on, which is why she insisted that our book be called *Communion*. Contact is not just about our learning new science and making new social and cultural discoveries. It is, more importantly, about this sharing of self. And incidentally, this has nothing to do with what is called possession. That's exactly what the visitors don't want to do.

For them, I don't think that anything is ever new. For us, everything is always new. They may know reality outside of time. We don't really know what the next second will bring. They hunger to share our sense of newness.

I think that the reason for this is explained by an insight that was published in the April 1977 issue of the magazine *Science*. D.B.H. Kuiper and Mark Morris made the observation that any intelligent entity appearing here from another world would have essentially nothing to gain from us except the results of our own independent thought. They would be after newness, and they would therefore be concerned about our state of preparedness to engage with them. As Kuiper and Morris speculate, "We believe that there is a critical phase in this. Before a certain threshold is reached, complete contact with a superior civilization (in which their store of knowledge is made available to us) would abort further development through a 'culture shock' effect. If we were contacted before we reached this threshold, instead of enriching the galactic store of knowledge we would merely absorb it." They continue, "By intervening in our natural progress now, members of an extraterrestrial society could easily extinguish the only resource on this planet that could be of any value to them."

Having been involved with them now for so many years —for much of my life, really—I feel that this is indeed the reason for their secrecy. But they now find themselves in a quandary: Our planet is failing so rapidly that if they continue to hide and wait for us to catch up, we might go extinct first, or enter into a period of chaos that will destroy what progress we have made, causing them even further delay.

Fortunately, it's not clear that our making more scientific progress is the only thing that holds them back. I think that what I might call psychospiritual progress is at least as important, and probably more so.

This would be why they have lavished so much more attention on this non-scientist spiritual seeker than they have on any scientist I am aware of. I learned, first through my Gurdjieff work and then by working directly with them, a grammar of communication that is both efficient, in the sense that progress is steady, and fruitful, in that the richness of communication is rapidly increasing.

The sensing exercise not only opens us to them and enables communication, it sends out a signal that communicates a good deal more than a laser pointer. When we place our attention on the nervous system, it glows in their level of reality like a little ember. Our dead can see this, too.

I came to understand this while I was at a conference hosted by William and Clare Henry in Nashville in September of 2015, just after Anne died. During a break in the conference, a woman came up to me, wanting to speak. She told me that she'd heard Anne say in her ear: "Tell Whitley that I can see him when he's sitting in the chair."

I understood at once what she meant. She had to be referring to the chair I sit in when I do the sensing exercise. My mind flashed back to something the visitors said back in about 1987 or 1988, when I asked them why they'd come: "We saw a glow." At the time, I thought that this meant the glow of cities, but in this moment, I understood that they'd been talking about the light emitted by the nervous system—light that I've known for a long time that they can see. And not only them. In the past, we could see it, too, as will become clear in Chapter Ten, when we discuss the oldest religious document in the world, the Pyramid Text in the Pyramid of Unas in Egypt, and what its extraordinary contents have to do with the modern experience of contact.

When I got home, I sat down and started doing it at once. A few days of this and my teachers soon arrived to assist, and soon I was in an almost miraculous relationship with Anne. It wasn't what's called channeling but rather a communication based on preparation that, unbeknownst to me, Anne had been making since she had a stroke in January of 2015.

It was at that point that she began insisting that I memorize W.B. Yeats' poem "Song of the Wandering Aengus." In that poem occurs the line, "When white moths were on the wing, and the moth-like stars were flickering out..." This proved to be the axis of all my future communication with her. I also remembered, after her death, that her favorite of my short stories is called "The White Moths," which is about a woman discovering that she has died.

What started to happen was that, when I was away from home and the surveillance system was running, when I talked about Anne, it would send me a text and I would see a white moth pass in front of the living room camera. This happened again and again, sometimes when I was lecturing about her at conferences, other times when I was talking about her. Each time, the moth passed in front of the camera and disappeared. No moth was ever found in the house, and no such moth ever appeared while I was there. It even appeared while I was at a banquet at the end of a conference about the afterlife and landed on one of the conferees' heads! It then flew off, and before the eyes of a room full of people who were watching it in amazement, simply disappeared into thin air.

It must be understood that contact with the visitors in all their strangeness and power is also contact with the part of humanity that is in a nonphysical form—that is to say, the dead. But this isn't only about a psychospiritual level of relationship. As we shall see, physical materials are involved and are being studied with astonishing results.

The deeper you penetrate into the experience, the harder to understand it becomes. It involves human and nonhuman beings in both physical and nonphysical states. It exists both inside and outside of time and space as we understand them and is larger than physical space and deeper than time. It is highly energetic and energizing. It is also challenging, difficult and can be dangerous.

It is hard. For example, when I talk about the sensing exercise, people usually want to know how long it takes to get some response. I can only relate what happened to me, which is that I started the exercise in 1970 and was initiated by the *Communion* experience fifteen years later, in 1985.

Because we are so focused on the material world and hardwired to see only what we understand of it, they have to fight their way into our lives. I think that they started trying to make me aware of their presence as soon as we moved into the cabin and they could see me doing my meditation in an isolated area, but one that was also near where they were engaged in some sort of enormous operation that was giving rise to hundreds of UFO sightings along the Hudson River. I had no idea that this was happening not thirty miles from the cabin, where I was spending inexplicably uneasy nights. The story has been ably covered by Dr. J. Allen Hynek, Philip Imbrogno and Bob Pratt in their book *Night Siege: The Hudson Valley UFO Sightings*.

From the time we bought the cabin, I was terrified that there was somebody lurking around outside at night. I installed alarm systems, bought guns and so forth—all of which is documented in *Communion* and other books. What is not documented is that in the summer of 1985, I began having terrible headaches, I think now because I was suppressing my awareness of the visitors. I say this not so much because of any specific scientific evidence that suppression like this causes headache but rather because I observed it happen at the cabin. Dr. John Gliedman, a dear friend and scientist who was there, along with a number of other witnesses, saw a very dramatic shaft of golden light come down from above. Alone among the ten people present, he could not make it out, even though it was right in front of him. Seeing it would have destroyed his world view, which he could not face. The result was that he soon found himself incapacitated with a migraine. Mine the previous summer had been so bad that I'd sought medical help.

By October, I think that the visitors were being compelled by my reticence to become more aggressive. If our relationship was going to grow, my refusal to admit to myself that they even existed had to be overcome. On the night of October 4th, we had our friends Annie Gottlieb and Jacques Sandulescu up for a visit. We dined at a restaurant and then returned to the cabin. We all went to bed almost immediately. Then, as I reported in *Communion*, "I was startled awake and saw, to my horror, that there was a distinct blue light being cast on the living room ceiling." (Our bedroom overlooked the living room, which had a cathedral ceiling.) I watched the light seem to creep down from above, then decided that the chimney of our wood stove must be on fire. "Then," I continued, "I fell into a deep sleep!" They didn't cause me to fall asleep. I did. I had to, or I would have had to face them. They must have waited for some time to see if I would rouse myself. When I didn't, they woke me up with a loud crack like a firecracker going off in my face. When I opened my eyes, I was "stunned to see that the entire house was surrounded by a glow that extended into the fog." (It was a very foggy night.)

The explosion caused Anne to cry out and our son, aged six at that time, to shout from his bedroom downstairs. Annie Gottlieb later reported that she heard feet "scurrying" across the floor of our bedroom upstairs, and Jacques observed light around the house so bright that he thought he'd overslept and that it was full morning.

I was then heard outside their door telling them that it was nothing, just go back to sleep. What had happened was that the light had gone out, meaning to me that there was no fire.

At the time, I thought I was reassuring everybody that all was well, but what I was really doing was suppressing what I knew perfectly well: that the visitors had come down from above and I had woken up to find the room full of them. When I leaped out of bed, probably right at them, they ran away. This had caused the scurrying that Annie Gottlieb heard.

In retrospect, I think that it was my noticing them even slightly that night which encouraged them to keep on trying to rouse me from the fixation the body causes on life in the time stream. Facing them means rising out of it, which forces the ego to face its mortality, which it is designed by nature to avoid at all costs.

On December 26th, my resistance was finally broken.

They are not our enemies. They are not our friends. I am all but sure that their purpose here is to become our teachers and in return to be rewarded with the sharing that they seek.

I find them challenging and demanding, but also exquisitely responsive to my needs as their student. They know the paths I need to travel but will never simply put me on them. Their skill as teachers comes in their ability to enable me to find my own way rather than show me what they think I should do. I know that I frustrate them. I've felt it and seen it. But no matter how poorly I do, I am rather sure that they will not give up on me unless I give up on myself. More than once in my life, I have rejected them and steered clear of them. But when I wanted to come back, they have always been there.

Each night during the first meditation, I bring to mind the children of the world. I live in a young neighborhood and see children every day. I see the wonder in their eyes, the joy, the hope. During the three o'clock meditation, I open myself to the visitors and in return receive knowledge and am exposed to their teaching. I think that I can feel, sometimes, that they are glad that this is happening between us. I know that they've staked a lot on it, because this is the only testament of its kind ever written down.

Communion is not just about contact between us and the visitors. It is also about the veil between the worlds falling, and we who are in bodies coming into real relationship with mankind unbound, that great, soaring wonder that we call "the dead."

When that happens, our understanding of life is going to refocus on how to live in such a way that we die with strong souls, which means without regret and not burdened by the memory of deeds that hurt others.

I didn't know it at the time, but as 1989 rolled around, I was about to enter a new stage of relationship with the visitors, which would open for me a door into the most enigmatic imaginable experiences and a whole new way of experiencing both life and afterlife.

In May of 1989, I was given a gift that at the time seemed like the most awful invasion of my body, my soul and my life that I could imagine. I was given my implant.

# 4

# THE IMPLANT MYSTERY

The spiritual experience of communicating with the visitors has such a powerful and pervasive physical component that I could as easily say that it is a physical experience that has a powerful spiritual component. The physical side of it—the technology—has left me with the most wonderful spiritual gift I can imagine possessing. While the implant in my left ear was put there in 1989, not until September of 2015 did I begin to learn how to use it.

What it does not do: It does not provide voice communication.

What it does: A slit opens up in my right eye's field of vision where I see words racing past. Also, it enhances research in a very unusual way.

On a warm May night in 1989, a man and a woman entered our cabin, bypassing the armed alarm system. They overpowered me and put a small disk-shaped object in the pinna, or upper part, of my left ear. I have described this experience in detail in *Confirmation*. After inserting the object, they left. A moment later, a flash of light filled the room, visible behind my closed eyes. There was crashing in the woods behind the house. The pressure released, and I immediately leapt up. I ran off in search of the intruders but found nobody and no sign of forced entry. As the alarm system had remained armed throughout, I decided that it must have been a vivid nightmare.

However, the next morning it became clear that the system had been tampered with, and later that day, my ear began to hurt, and we noticed that a lump had appeared in it. (In *Confirmation,* this event is dated as 1994. May of 1989 is the correct date.)

At some point over the next few days, the thing turned on, making a growling-whining noise in my head and causing the ear to turn bright red. I was absolutely terrified. Frantic. I wanted to cut my ear off. Anne tried to calm me down. She thought it was a gift. I felt like I was being tracked. She said, "The visitors don't care if you go down to the store to get some beans." I had to admit that she had a point. I treasure my privacy, but I don't do or think anything that I fear being known by others. Still, I wanted the thing out.

The close encounter experience is full of tales of implants, and my interest in them was obviously intense. In the early 1990s, I met Dr. Roger Leir, who was organizing implant removals in California. Anne absolutely forbade me to let him remove mine. Hoping to soften her attitude, I made sure she attended one of his very professionally conducted removal surgeries with me. It was an awesome and moving experience to see the object come out of a woman's calf muscle. It wasn't much larger than a pumpkin seed and gleamed because, as I later learned, it was encased in epidermis. I have discussed the early implant work extensively in *Confirmation*, published in 1999, but it was not until 2015—almost immediately after Anne died—that I began to be able to understand how to use mine. Just recently, in September of 2019, I have had a CAT scan done of it and have had a visit from two people who explained more about it to me, as I will relate shortly.

But don't expect anything straightforward.

Dr. Leir offered to arrange an extraction, but Anne felt strongly that I should try to understand it first. So I delayed. By that time, I had learned to have the greatest respect for the extraordinary role she was playing in our relationship with the visitors. The implant really troubled me, though. I felt watched. There was a distressing sense of being trapped in my own body with the thing. It was claustrophobic.

When we moved to San Antonio in the mid-nineties, Catherine Cooke, then president of the Mind-Science Foundation, introduced me to the head of materials science at Southwest Research Institute in San Antonio, Dr. William Mallow. When I told him about what Dr. Leir was doing, he was eager to learn more about the objects. I called Dr. Leir and arranged to obtain a group of samples, which he brought personally from Los Angeles.

While we were not allowed to conduct any work with the official blessing of the institute, we had free use of its equipment. Under the scanning electron microscope, we found the objects to almost all be meteoric nickel-iron and, except in one case, to be unremarkable except for the fact that, when *in situ,* they were usually encased in a capsule made from the host's epidermis. As the body does not have the genetic encoding to generate epidermis inside muscle, the encapsulations had to have been created artificially. In many cases, a small scoop-mark scar could be found on the body of the host individual. This scar would have removed a bit of epidermis from beneath the layer of tissue on the surface, the stratum corneum.

One of the objects we received was quite unusual. It had been emitting a low-level FM signal when it was still sited in the host. Under the scanning electron microscope, it appeared to be another fragment of nickel-iron. As it had been broadcasting, we decided to see if we could detect any crystalline structures using an X-ray diffraction machine that was available at the nearby University of Texas at San Antonio campus. The first pass returned a typical response for nickel-iron, but subsequent passes showed no return at all. A check back on the SEM indicated that the fragment's composition was still the same: nickel-iron. We returned it to X-ray diffraction, only to find that it never again returned a signal. Upon first being touched by X-rays, it had become X-ray invisible.

So obviously, at least in that one case, we were looking at something with unknown characteristics.

Working on the implants, I was constantly aware of the one in my ear. Bill knew about it, of course. He was itching to get his hands on it, and I was just as eager to get rid of it. And yet, some of the people who'd had theirs removed in Dr. Leir's program told me that they felt a real sense of loss afterward, almost as if a friend had died. Others were glad to get rid of them. And I had Anne advocating for it, of course.

Then we met Dr. John Lerma through mutual friends. As he wasn't involved in the UFO world, Anne was more comfortable having him at least take a look at it, so I made an appointment. As he examined it, he said that it looked like a small cyst and that he could easily remove it. All that would be involved would be a brief office procedure.

I was excited, frankly. Anne was not. In fact, she insisted that I promise to try only the one time, and if it didn't work, to leave it in. I could see in her eyes something rather fierce. I knew how intense my brilliant wife could be when she thought I was making a mistake. I offered to cancel the surgery, but she said, "No, you have to do it. It's been bothering you for years."

So, on October 9, 1997, we appeared at Dr. Lerma's

office, camera in hand. I wanted to videotape the whole procedure, and I'm glad we did. Anne was extremely nervous and laughing all the way through, very typical of her when she was uneasy about something.

Dr. Lerma anesthetized the ear and, as Anne taped, made his incision. He said, "It's a white disk." Then he touched it with his scalpel...and it moved away. He was shocked, because it had been fixed tightly under the skin. And yet it had now moved. All he had were two small fragments on the edge of his scalpel, which he deposited in a specimen container. He then withdrew. I took the fragments to Dr. Mallow, who examined them under the scanning electron microscope. One of the samples was just cartilage, but the other one contained crystals of either calcium carbonate or calcium phosphate.

A lab technician, who had part of the fragment that Dr. Lerma had removed, called him and asked if it was a practical joke. Under the microscope, he was seeing proteinaceous material that was adhered to a metallic base. He told Dr. Lerma that, as far as he was concerned, it was a piece of technology. Dr. Mallow then called me and confirmed that this was also his thought.

Two days after the surgery attempt, the object moved back up into the top of my ear, where it remained until three weeks ago, at which time it moved about half an inch down along the pinna. I described the situation to a neurologist who is associated with a nongovernmental but related program that studies these objects and the people who host them, and he recommended a CT scan, which I got on September 23, 2019. What happened was surprising. The object no longer appears to be there. And yet, it still works. In fact, by the time it *started* working in 2015, it might well have already been gone, at least when it comes to metallic parts. I can still palpate it, but what I am feeling is just the calcium that my body surrounded it with, not the object itself. I suppose it's possible that it is X-ray invisible, but a 3D scan failed to turn up any gaps or hollows, so that seems unlikely.

Gone or not, it still works, and I consider it the most valuable thing in my possession.

The reason has nothing to do with visitors, though.

Between the mid-nineties and 2015, it was mostly just there. Once in a while, it would turn on, causing my ear to become red and hot. This would happen most consistently when I was with other people who'd had the close encounter experience. It happened once at Southwest Research, and a signal was detected in their signals lab, but I was told that I could not be given the details. The most dramatic event came in April of 2016 when I was rehearsing the first conference speech I had given in many years before a relative who has also had contact experiences. She said, "Your ear's turned bright red." I laughed and explained that I probably had a larger audience than one.

It was in September of 2015 that I had noticed a dramatic change that seemed related to it. I happened to be in a room with a white wall that the sun was shining on, when I realized that I could see a neat oblong slit in my right eye. It was filled with movement. When I concentrated, I could see words racing past, but too fast to read more than the fact that they appeared to be typing. It was perhaps the most familiar of all fonts: Courier.

I peered at the speeding words in disbelief, then awe. This was clearly not a hallucination—or if it was one, it was, to say the least, unique. It appeared to me to be that technology was being applied. When I did manage to read an occasional word, they seemed random. Over time, though, I realized that this was not the case. The words were related to what I was writing, but not directly. For example, I started writing a historical novel, *In Hitler's House*. When publishers saw that it was written by Whitley Strieber, nobody would buy it. This isn't only because of prejudice, of which there is undoubtedly some, but due primarily to the fact that my sales have been weak for years. I published it myself under the pseudonym of Jonathan White Lane.

What I found, as I worked on the book, was that the words flashing past would increase the richness of my associative process. If I thought, for example, about the way a certain character might react to an insult and came up with the word "arrogance," words like "ignorant," "fire," "loneliness," "childhood" would be speeding past. The way that they were and are indirectly related to my thoughts enriches the associative process and adds depth to my writing. But it does more than provide enhanced association like this. It is also a marvelous research tool.

The novel is a faux memoir by a young German American who innocently falls in with Hitler in 1931, becomes an allied spy in 1935, and stays close to Hitler until the end of the war. To write it, I needed to know details about life in the 1930s and life with Hitler that were so true-to-life that they would seem to have been written by somebody who was actually there. I found that the implant would respond to direct questions but in quite a unique way. For example, I asked it, "What kind of toothpaste did Hitler use?" A day later, even though I was looking in Google English, it turned to German, where I found a reference to a book written by one of Hitler's valets. I bought the book, which was in German. As it happened, the next night I was with somebody who spoke fluent German and was willing to translate it for me.

If this had happened only once, I would think that it was a coincidence. But, like the associative enrichment, it happens with such consistency that I feel it is part of the way the implant works. It has also provided me with unique insights, such as the material on the social consequences of overpopulation, the relevance of the mystery of the fine-structure constant to understanding the nature of reality that will figure later in this book and many, many other things. In fact, I can say that this book has two authors: me and my implant. As to who might be doing all that work, I once asked it, "Who are you?" The reply came back at once, and slowly enough to read clearly: "It's me, Anne."

I remembered Anne's gentle but persistent opposition to my having it removed, and I have to say, at that moment, I thanked her from the depths of my soul that she had convinced me not to do it. As to whether or not she was conscious at the time of what it would eventually be used for and who would use it, I cannot be sure. But one thing is not in question: whatever condition it is in, it is an extremely useful tool. I have also recently learned, under some truly wonderful circumstances, much more about its origin, functionality and who installed it.

As things stand now, I would never allow it to be removed. I use it all the time. It is constantly leading me in new directions in my work, answering unanswerable questions and enriching my creative process.

Just recently, for example, I tested it by asking to be told something that would be important to this book, but which I knew nothing whatsoever about. Within hours, I found myself looking on Google at material about something called the fine-structure constant. I was looking for something else, and I cannot say quite why that particular phrase came up. I'm sure it's explainable, but it was not an expected result in the search I was doing, which was about the psychiatrist Carl Jung, and was for information I needed for a reading group I attend where we had been studying his *Red Book*.

I can't even be certain that the implant was at work here, but it seems as if it was, because what I have learned about the fine-structure constant, which turns out to be one of the great mysteries of physics, is directly relevant to the new vision of reality that is central to this book.

The fragment of the implant that was captured by Dr. Lerma consisted of a metallic base with cilia attached to it. They were motile, that is to say, as the lab technician told the doctor, "alive." It thus appeared to be a piece of biotechnology. Like many of the implants that Dr. Leir extracted, it had the ability to move away from the scalpel. Dr. Lerma was not expecting this, and his surprise is evident on the videotape. I very well remember how my ear burned when it returned from the earlobe to the pinna.

At about 4 in the morning on Wednesday, September 18, 2019, a stunning event took place that explained a great deal more about the implant. Since I experienced the movement of the implant I'd been feeling a paresthesia in the area, I had become more and more nervous about it. During my mediations in the nights leading up to the 18$^{th}$, I was complaining and saying that I was going to get it removed if the sensations didn't stop. (They were later diagnosed as a pinched nerve in my neck.)

On the 17$^{th}$, I did my usual 11 PM meditation, then at 3, the second one. I was in bed, just getting to sleep when I heard a soft knock on the door. I looked at the clock. It was 3:44. I got up and went to the door. I have heard these knocks many times and generally peer out the peephole before opening the door, always to find nobody there.

This time, I swung the door open and there stood two men. Before I could so much as gasp in surprise, I felt a change come over me that was akin to the twilight sleep one might be given during a minor surgery. It was not so powerful that I couldn't walk or move or talk, but I was definitely no longer in a normal state.

Incredibly, I recognized one of the two young men. I had last seen him when he was about twelve. He had been with his two sisters and an individual who may have been from the Department of Defense. They were special children with capabilities that most of us do not possess, primarily an ability to read minds. When I saw him when he looked about twelve, it was 1996, so he was now in his mid-thirties.

I wish I could say more about the circumstances under which I saw him then, but there is very little more to say. The children were introduced to me in a public space and immediately began talking to me in my mind while the adult they were with watched me with twinkling eyes and we all laughed with delight. Because it was delightful. They were delightful. I was just thrilled to see that there were human kids with this capability. I don't know how many there are, but I hope many, and that many more are being born. I have no idea how it works, or even theoretically how it might, but it is a wonderful thing and a real advance in human evolution. May these kids thrive and may their tribe increase!

Anyway, I recognized him at once, and I was absolutely amazed to see him. He gave no indication that he recognized me, and by that point, I wasn't capable of speaking or even moving very much. They were being very careful with me, and with good reason. I would damn well take a picture, steal an artifact—anything I could manage—and anybody from that side who works with me must know that perfectly well. They sure did.

In any case, they had a small portable typewriter with them. It looked like something that was commonplace before computers. I was told that it was what was used to generate the words that race past in the slit in my eye. I looked down at it. He put it in my hands. I said that I didn't see any sort of radio or anything. It was just an old typewriter. Very trim and surprisingly light.

He then explained that the words I see aren't generated outside of my mind but are drawn up from deep in my unconscious. When they are typed, they appear in the slit. Thus, they are drawn from a level of my mind that I cannot reach to the edge of consciousness where I can make use of them.

I asked how in the world that might work without any communications device. He explained to me that it was in the typewriter's platen. So I asked again how it worked. He said that he didn't know but that it had been developed by a Dr. Raudive.

This name was vaguely familiar to me. When I Googled it the next day, I found that this was a Dr. Konstantin Raudive, who had been a colleague of Carl Jung and who had worked for years on what is known as EVP, or electronic voice phenomena. This involves the design of devices that enable people on the other side of the barrier between the living and the dead to communicate. After his own death, individuals using EVP found that they could communicate sporadically with him. He was continuing to work on the creation of this technology from the other side.

In fact, the only other person I know who has the slit open up in his visual field that has the words racing through it is a man who has studied EVP for most of his adult life and is an expert on the work of Dr. Konstantin Raudive. He reports no evidence of anything in his body suggesting an implant. (Of course, I discovered this the day after I learned about Dr. Raudive, another of the strange coincidences that fall like rain on a clear day when the implant is involved.)

The two men explained that it had been repositioned because it was stressing my right eye. The intraocular lens in this eye, the membrane behind it, and the retina have all been affected by calcium deposits over the past few years when I have been using the implant almost constantly. (An intraocular lens is a replacement lens that is used to correct cataracts.)

They then asked me if I still intended to have the implant

taken out. As the IOC can be replaced and the membrane removed and the retina is not symptomatic, I said that I would not. They then left. I stood there staring at the door. My mind was racing. I was still in twilight sleep and had to move very carefully until it wore off a bit. I tried to go to my couch and get back into the sensing exercise, but I could not manage it. I was exhausted and instead fell into bed and into a deep sleep.

A few days later, the study of the CT scan came back: The implant they didn't want me to take out isn't there.

Or is it? I wonder what would happen if I had the calcium deposit that remains removed. Or would it race off to some other part of my ear like the metallic object did in 1997?

One thing is sure: If you don't like mysteries, especially unsolvable ones, stay away from the close encounter experience.

I no longer think that the implant has anything much to do with nonhumans. The fact that I didn't learn to use it until after Anne passed away, and that it continues what she did when she was alive, which was to be a fabulously brilliant muse to this struggling scribbler, and what I learned on the morning of the 18th has convinced me that it is a communicator between the living and the dead.

I hope, of course, that more people will gain access to such devices. However, I also think that there are liable to be abuses, and I wish to say that my implant has never provided channeled information. As I said at the beginning of the chapter, there are no voices involved. Rather, it does just two things: send the words flying past in my visual field and create synchronicities that support my research.

I reach up and feel it. There it is, quietly doing what it does. A strange event that lives with me twenty-four hours a day. I turn a lamp against the wall, then sit watching the slit. As usual, words race past.

Finally, I catch one. It is "harmony."

# 5

# THE FIELDS OF ASPHODEL

During the weekend of July 19–22, 2019, I went to a place of great human suffering and incredible power. While there, I had extended, days-long access to another world, an experience that went far beyond anything else that has happened in this lifetime of strange and extraordinary experiences. I think that what happened offers a major clue about the origin of the visitors, and possibly also of their enigmatic human allies.

I had been invited to a small conference at the All Nations Gathering Center on the Lakota Sioux Pine Ridge Reservation in South Dakota, where I was to give a talk about *The Afterlife Revolution*. It was hosted by Dallas Chief Eagle and his wife Becky and organized by Mia Feroleto, publisher of *New Observations Magazine*.

Before going, I had learned some of the reservation's history, but it had offered no clue about what was actually going to happen to me there. Like most people outside of American Indian culture, my awareness of the spiritual power of their religions was very limited. Being a Texas German, I was aware that my ancestors had a high opinion of their religion and spiritual development. Why, I did not know. I do now.

I also knew that Pine Ridge was the site of the 1890 Wounded Knee Massacre and the Wounded Knee occupation of 1973. On December 29, 1890, the US Army had opened fire on a group of 300 Lakota Sioux, killing 90 men and 200 women and children. In 1973, Wounded Knee was occupied by 200 Oglala Lakota and members of the American Indian Movement in protest over corruption in the tribe's government. This led to a siege that lasted two months that left two Lakota killed and fourteen wounded and two federal officers injured. I also learned that Oglala Lakota County was the poorest county in the United States, with an average annual income per person of just over $8,000. Officially, the average life expectancy on the Pine Ridge reservation is 66.81 years, but statistics attributed to the Pine Ridge hospital cite a life expectancy among women of 55 years and men 47 years. Suicide rates are high, especially among teens, driven by the sense of hopelessness that infects their lives like a virus. During the winter of 2015–2016, one 12-year-old girl killed herself because her family could not afford heat, and she could no longer bear the cold. Alcoholism affects 85% of the population. Drug abuse and crime are rampant, and living conditions are dreadful beyond anything I have ever seen in my life.

None of this is an accident or due to laziness or any such issue. It is because of the location. During the 19th century American Indian wars, the Lakota Sioux were intentionally confined to this place because it is so lacking in resources. Distances are long, so work off the reservation isn't economical for most residents. Because of its isolation, lack of good farmland and general scarcity of exploitable resources, there are few jobs on it, contributing to a chronically high unemployment rate.

While I found an oppressed people there, I also found that it was a place of great human spiritual power, in fact, power beyond anything I have ever known anywhere. I have some idea of what this power is, which I will discuss in depth in a later chapter. I had not been on the reservation for more than a few hours before I began to feel it. And when I say feel, I am not talking about something vague—some sense of unusual energies. Far from it.

On my first morning there, when I happened to close my eyes during a drive of half an hour or so, I saw movement behind my closed lids—what looked like shadowy trees and rolling hills, but not the ones we were passing. Surprised, I opened them immediately. I couldn't understand why I'd been seeing anything at all. When I closed them again, what I saw simply took my breath away. I sat there watching an entire second landscape flow past the car. Although it seemed to be twilit rather than sunny, the effect was so vivid it was like wearing a virtual reality headset.

I was flooded with strong, poignant and yet contradictory emotions. There was at once a sense of homecoming and homesickness. It wasn't as if I was in two places at once, but rather looking out the windows of my heart into two worlds that have been locked forever in a secret embrace and seeing that wonderful, sweet thing for the first time.

As we drove along, I sang out the different features I was seeing. "There's a creek over there, we're passing under an arbor of trees, there are long hills on the horizon. Oops, the road's gone off down the hill." Among those in the car who heard me doing this was our very kind driver, Kevin Briggs, who unfortunately could not close his eyes and look as the others did. Conferees Alan Steinfeld, Ananda Bosman, Mia Feroleto and others did close their eyes. Some saw it vaguely, others not at all. Only Ananda and I saw it clearly.

Even though the image was shadowy, it was extremely

detailed. I could pick out individual trees, fields, even a narrower version of the road we were on. After a few moments, I realized that I was watching not another world altogether, but another version of the landscape we were passing through. It was a bit more rough, with occasional gorges and generally wider streams. The other road was not only narrower, it wasn't graded. The result of this was that it sometimes wound off down a hill while we continued along the graded version in our world. This would leave me with the uncanny sensation that the car had taken flight.

The vision didn't go on for just a few minutes, but for the entire time I was on the reservation It continued whether I was riding in a car, walking, sitting or standing. For those three days, I was living in two landscapes at once. After I closed my eyes, it would take about thirty seconds for the other world to appear, but it did so reliably. When I was standing somewhere, I could look down and see grass and gravel that was not present in this world. I could bend down and look closely, even to the point of being able to count the number of petals on flowers and observe the details of grasses and the discolorations on stones. I could touch and smell nothing of the other world. In this sense, it was very much like out of body travel, which detaches you from those senses. I was not physical in that world, and I have to wonder if that might not be how our visitors experience this one. I tried using the sensing exercise as a tool for physically moving into the other world, but it didn't work. Nevertheless, it is my strong sense that what we think of as technology is not what enables things like this to happen. I think that it's something to do with attention, concentration and the brain, and possibly also requires the cooperation of an outside energy that is itself conscious. My thought is that my lifetime of doing the sensing exercise and the changes in my brain that have resulted have made me more able to see this other universe and, to a limited extent (so far), interact with it.

The changes I am referring to involve a brain area called the dorsal striatum. It contains two regions, the caudate and the putamen, which are connected by an area of white matter called the internal capsule. There is a study under way that suggests that the density of the white matter region may govern the degree to which an individual possesses intuitive sensitivities.

I have thought that meditation might somehow increase this density. The authors of the as-yet unpublished study don't see any indication of this so far, but another study, published in *Psychiatry Research* in January of 2011 and entitled "Mindfulness practice leads to increases in regional gray matter density" states, "Analyses in *a priori* regions of interest confirmed increases in gray matter concentration within the left hippocampus. Whole brain analyses identified increases in the posterior cingulate cortex, the temporo-parietal junction, and the cerebellum in the study group compared with the controls."

If the dorsal striatum is also affected by meditation, it might explain why, over years of doing it, people seem to become so much more intuitive, as has happened in my case. My own brain has been observed to have a very dense internal capsule between the caudate and the putamen, but there is no premeditation MRI to compare with the one that is available. I do feel that fifty years of meditation has made changes that involve opening my mind to new areas of vision and new ways of seeing.

However, I would not ignore the power of the heart, either. While we now attribute thought and feeling exclusively to the brain, I think that the older vision of the body, with the heart as the emotional center, should not be dismissed. I say this because the whole experience was so emotionally powerful, and like we do so many emotions, I felt it in my heart, and it is in both head and heart that I carry it now.

There is energy involved, which I think is conscious and capable of deciding exactly what it wants and does not want to do. A big part of coming into communion with the visitors, I feel, is opening ourselves to the wishes of this energy and attempting to understand what it might need from us and that will help it fulfill its great aims for the unfolding of space-time.

The mindfulness practice researchers found that the changes they detected came about rather quickly, after only eight weeks of meditation, so perhaps building a brain that is receptive to contact is not all that lengthy a process. The longer time, I think, would involve waiting for one's little glow to be noticed. My experience of this is that determination counts. You have to want it a lot and work hard for it before anyone will show up to work with you. To get darker, more exploitative aspects to appear, such as happened to Matt, all that is needed, it seems to me, is curiosity. Lack of preparation, though, is clearly not a good idea.

That preparation should include reading material about what it is like to live with the visitors. Authors such as John Mack and Kathleen Marden offer carefully researched texts, and Anne Strieber's *Communion Letters* is a treasury of personal accounts. I would avoid channeled material, as there is never any way to know where it actually comes from, an outside source or the author's imagination.

Beyond that, later in this book, I will explore the importance of coming to this with a strong, healthy soul. This conscious energy, as I have experienced it, is very reflective. If you feel fear, it will be fearful. If you conceal guilts, it will look right into you and at them. Having been face to face with the grays when I was still an unregarded soul, I can assure you that this is an incredible shock. Those glittering pop eyes burned into me. It was like facing goblins, and I felt like some deep part of me was about to be devoured. The ancient Greek aphorism "know thyself" is of critical and foundational importance to establishing a relationship with the visitors and all that comes with them. Later, I will go into some detail about how one uses it to build the sort of strong soul that will have no reason to find those penetrating eyes frightening.

I am purposely being a little vague here about who I mean —is it the strangely formed visitors I'm talking about, our own dead, or a sort of field of disembodied consciousness?

I don't think that it's useful to make such differentiations. Whatever aspect of it comes into contact with you, so does it all. Best to think of it as a vast field where different sorts of flowers grow, some of them appearing one way, others another way. No matter which way you go, you remain in the field.

A specific event that causes me to suspect that we are dealing not only with specific entities but a conscious field took place at the Contact in the Desert Conference on June 2, 2019. As I was heading to a lecture by Dr. Jacques Vallee, I noticed an odd change in the atmosphere. It was as if the air pressure had dropped. My ears popped. Sound faded. I said something to the person beside me, who acted rather strangely. He seemed to be pretending that I wasn't there. I thought perhaps he disapproved of me in some way. I went on to the lecture, and by the time I arrived, the sensation had faded.

Some weeks later, an attendee at the conference wrote me, "I was on my way to Jacques Vallee's lecture when I saw Whitley coming down a side path in my direction, looking deep in thought. Our paths were totally going to cross. Just before we were about to intersect, I glanced down to check that I had my phone in my bag. A bright flash from Whitley's direction caught the corner of my eye. Oddly, I thought, 'That's him flashing to another dimension.' In the same instant, I looked up—sure enough, no Whitley. He was there, and then he wasn't. It's hard to describe how confusing and odd it felt. The atmosphere suddenly felt heightened, and the sound seemed to drop away. At the same time, a sense of immense great benevolence came over me, as if someone with kind good humor was reassuring me that everything was ok. It felt like a gift, meant for me. It felt like magic." She then spent some time looking for me without success and moved on to the lecture. She and a friend were taking seats "when the next person entered. It was Whitley Strieber!" For my part, I only felt the change in atmospheric pressure that also affected her. I had no sense of moving through another universe, but perhaps when this happens, we leave any memories we have gathered there behind when we return. This would be one explanation for the ubiquitous experience known as missing time that is reported by close encounter witnesses. We are not being taken aboard spaceships at all but moved into the companion universe, and our memories of events there return with us only in fragmented or suppressed form, or not at all.

It is important, I feel, to note her comment that "a sense of immense great benevolence came over me, as if someone with kind good humor was reassuring me that everything was ok." I think that this may have been a moment of direct communication with the conscious field itself, rather than specific entities that are part of it. When I am in touch with it, there is always a sense of joy, even hilarity. Those moments never fail to remind me of Anne's love of the 14th century mystic Meister Eckhart and his statement that "God laughs and plays," and of her own central teaching: "Have joy." She adopted this because it was one of the very, very few things that the visitors have ever said to me in ordinary language.

She felt that it was what lay beyond our suspicion and fear, just out of reach.

I think that my vision was opened to this other world at Pine Ridge by the action of conscious energy, not specific entities. I will say this: I have never had more fun in my life than I did while my mind was open to this vision. It was so fascinating, so tantalizing, so extremely interesting. It has absolutely inspired my curiosity and made me want to somehow walk in those fields. Maybe I'd be unwelcome to the inhabitants or even devoured by something, but maybe also by passing through the wall between our worlds, I might make a door that others could enter. I have been asking the energy for a chance to give it a try.

Having asked many times in my life for impossible things to happen and seeing them proceed to unfold (always to my great astonishment), I don't think that this entirely unreasonable and absurd request is at all impossible to fulfill. We shall see.

The other world was just as complex as this one, with streams, trees, fields, gorges, grasses and flowers and a sky complex with flowing clouds. In general, though, I didn't see many structures, and no people. From time to time, I'd see a house or cabin. Sometimes a white square would flash on the distance then quickly blink out. Was that a person, perhaps distorted by some quirk of consciousness that we don't yet understand?

When I looked up with my eyes closed in the car, I saw the sky and passing clouds. When I opened them, there was the ceiling. The weather in the other reality was similar but not the same. It seemed more unsettled. There were storm clouds there that were not present here. The moon was waning and gibbous in both realities, but in the other seemed to me to be a bit less gibbous, as if it had been full starting on about the 13th rather than the 16th, which was true here. Over the weekend that I was on the reservation, it was rising late at night, but when I closed my eyes, it was well risen in the other reality by about half past nine. Even stranger, when I opened them, for a few moments there was a sort of hazy glow where the moon had been in the other world. This would slowly fade into the normal night sky. I wondered, then, how close I was to slipping into the other reality.

On my second day on the reservation, I had the privilege and honor of being allowed to witness an hour of a private family ritual that I found to be among the most sacred things I have ever experienced. It involves chanting and drumming and dancing. It was deeply moving to me.

The dancers fast and dance over a period of days. In the hour I was allowed to be there, I danced as well, entering the ceremony as best I could. The chanting stirred my heart and my soul, the drums shook my blood. When I closed my eyes, the area where the ritual I was watching was taking place became an empty meadow. I could still hear the drumming, but it seemed to now be coming from the right, not the left. When I looked to my right, where the new source seemed to be coming from, I could see the edge of a low hill. The sound seemed to be below it. When I opened my eyes, the hill was no longer there, the meadow was again filled with dancers and the only drumming was coming from the left.

At one point, I noticed people looking up and pointing. But what could it mean? How could they not pay attention to such an event as this? Still, I was curious. I looked up, too, and there at the top of the clear blue sky was a small object. It was light tan in color and seemed quite high. Was it a balloon? I watched it for a few minutes, but it didn't go anywhere or do anything—just hung there, motionless and silent.

I kept my feelings inside, but I did think that this was the visitors. I can't say that I felt their presence, which has sometimes happened in the past when I have seen their devices, but my initial reaction was that they were there to honor the ritual. They would have known that I would write about it.

I asked around about the object. Some of the people thought it might be an FBI drone, but others said that they were seeing beams and little balls of light coming out of it. However, the FBI keeps a close watch on these people. The Wounded Knee Occupation of 1973 was viewed as an insurrection against the United States. So maybe the FBI was watching and maybe they would have used a drone. As for me, I watched the object off and on for a total of about fifteen minutes but did not see any unusual phenomena associated with it, except for the fact that it was motionless to the point of being uncanny. The eye expects things in the sky to move, however slightly. This did not. I was reminded of the UFO footage taken by fighters from the carrier Nimitz in 2004 and released in 2017. The objects on those videos are not aerodynamic but held aloft in some other way. This object looked exactly like that. Others saw similar objects at various other times during the conference, so my thought here was that this was indeed the visitors. In September of 2019, the US Navy admitted that the objects recorded by the Nimitz fighters were indeed unknowns.

The one remained motionless for too long to be a balloon. Even if it was motionless only for the time I observed it, that would be too long for one. There were balloons being released as part of an experiment in Sioux Falls, but that city is to the east of Pine Ridge, and the winds that day were out of the west.

When I left, it was still there, still motionless. Individuals who left after I did confirmed that it was still in the sky at that time. When it did finally depart, it moved away slowly and was seen again over other parts of the reservation.

That is also quite a long time for a drone to be hovering

without being returned for a recharge. When they hover, they aren't absolutely still. Neither are they completely silent. I have been unable to find any commercially available drones that can remain aloft more than two hours. This was the limit, as of the summer of 2019, of the longest duration drone, the HYBRiX.20.

Still, although I do believe that the object was an unknown, I can't rule out the possibility that the FBI has silent, long duration drones that can hover for hours. I don't believe it, but I can't rule it out.

This was all pretty strange, even for me. Strange and wonderful. What happened next, though, was even stranger. Led by a member of the American Indian Movement, who was also one of the last direct descendants of somebody who had survived the 1898 massacre, a group of us went to the Wounded Knee memorial on the reservation. As I stood looking down at some of the graves, another member of the conference stood beside me, also looking down at them. I was vaguely aware that he had moved away then returned. I thought nothing of it until later when he came up to me and explained that, when he stood beside me, he could see down into the graves and could see the broken skeletons that were lying there in the earth.

As he explained this to me, I could see the puzzlement on his face. I don't blame him, as I have never heard of anything like this, not in all the literature of high strangeness that I have read in my life. There are a few cases where people were said to be able to see through objects. In Greek myth, Lynceus of the Argonauts was supposedly able to see though walls and into the ground. But nowhere is there a story of somebody who could somehow confer this power on another person while not possessing it themselves. The visitors can pass through walls. When out of the body, so can we. But X-ray vision, especially X-ray vision by proxy—no, I believe that this may be the only such story that has ever been told. I think that all three stories, though—seeing into the other world, apparently popping into another dimension and conferring this power on the person standing beside me—all have to do with the presence of the same energy. I have to say, though, even as I write this, I can see those broken skeletons in my mind's eye and feel the cruelty and shame of the massacre.

When we went to the nearby Badlands one evening, I found myself still able to see the other world there, too. I immediately noticed that hills were not as dramatic. This would mean that there has not been as much erosion there as here. When I looked down at the ground, I could see more grasses there than here, also. So the geologic history of the other world might be less violent, but at the same time, it also appears to be less populated. I say this because the roads on the reservation in this world are graded and in the other world they are not. I kept hoping that I would see a vehicle, but I never did. Had I done so, I wouldn't be surprised to find that it was horse drawn. This is because the roads in the other world were sometimes unpaved, and the tracks in them were narrow, suggesting wagon wheels. The paved areas were black like macadam.

As I left the reservation on Monday, I also left the other world behind. By the time we were twenty miles from Rapid City, I could no longer see it. On the plane back to Los Angeles, I gazed out the window at the gentle landscape far below and thought long thoughts about my strange life and the strangeness of life in general, as it was being lived in the cities and towns we were passing over. I had left the greatest mystery of my life behind. I'd be a fool, though, if I didn't live a "never say never" life, so for all I know, it's going to return to visibility sometime. I regretted not being able to enter it. To do that, however, I knew that I was going to need to find out more.

There are many instances in literature of people encountering other realities. Matt and his mother saw a field from the Pleistocene at the end of his runway. In his book, *Hunt for the Skinwalker*, Dr. Colm Kelleher describes an event that took place on a mysterious ranch that had been bought for study because of all the paranormal activity taking place there. The purchase, made by the government, was managed through Robert Bigelow's foundation. On the afternoon that the scientists who were to take it over arrived, a remarkable event took place. Just before they arrived, a huge wolf came up out of a marsh and attacked some goats in a corral. The ranchers drove it off by shooting it, although the bullets did not seem to penetrate or draw blood. It leaped out of the corral and ran back into the marsh.

When the scientists arrived, they found its footprints, which led to the center of the marsh and then disappeared. Casts taken enabled them to estimate the weight of this ghost animal at 300 pounds. When I read this, I immediately realized that this was probably a dire wolf, another Pleistocene creature.

So is the other world in the past? I have no way of knowing that, but I do suspect from all I saw and all I have learned about it that it does not have the level of population or development that our world does, meaning that, in it, those creatures might not have gone extinct.

On the night I got home, I sat as I always do at 11 and took my attention to my physical sensation, letting my mind wander free. I asked to understand the place I had seen. I asked if it had a name. I never do more questioning than that. My experience is that I don't need to beg, pray, do rituals or anything like that. What is important is to be prepared for a response that might happen very quickly and—above all—in some way that is liable to be quite asymmetric.

Nothing happened for a few days, and then on August 3rd, I had three dreams. They were what are known as lucid dreams. I've had a few of these in my life, but I don't recall any that were as vivid as these.

In the first one, I was standing outside the All Nations Center at Pine Ridge with my eyes closed, looking at a beautifully curved wooden fence that had been one of the most striking features I'd seen during my time observing the other world. Three Indians came up to this fence. I could tell from their clothing that they were probably from the ritual I'd heard taking place in the other world. They looked human but had differently shaped faces than we do, and their eyes were somehow different. Was this because they were structurally mirror images? I don't know how to asses that. All I can do is report what I saw.

They told me that, because this writing is going to help Pine Ridge in some way, I had been given my vision as a gift —which is all well and good, but I'd like to know who gave the gift and what had to be done to deliver it. And above all, what can we do to initiate contact from this direction?

The next thing I knew, I had two more dreams in rapid succession. They involved two of the few people left alive who were indirect witnesses to the Roswell event and know details of the bodies.

One of them said "Strieber-Greek" and warned me to be careful.

When I woke up, I wondered what in the world that might mean. I have nothing to do with Greece. Greeks play no role in my books. I am not Greek, speak no Greek and only know one or two Greeks, and them not very well. So I Googled the phrase—and got quite a surprise. There was a "Professor Strieber" mentioned in a science fiction novel called *Uncle*

*Ovid's Exercise Book* by Don Webb that was published in 1988. The passage in which my name appears contains the sentence, "The Greeks placed the coins in the mouths of their dead that they might pay Charon to ferry them over the Styx into the gray fields of Asphodel in the interior of the Earth."

Asphodel in Homer's *Odyssey* is the abode of the dead, twilit and leached of color. This is why the familiar little gray meadow flower one sees in some parts of Europe is called the asphodel. At first, I thought, "Oh, dear, perhaps I was being warned that I'm destined for some dreary land of the dead." Then I was glad that the sensing exercise hadn't worked to get passage!

However, Homer doesn't only condemn Asphodel as the land of the unwanted dead. A less known passage in the *Iliad* describes it as a place fragrant with lovely flowers.

And indeed, in my third dream on that night, I saw the most gorgeous field of blue flowers I have ever beheld. In fact, it was the most enchanting shade of blue I can imagine —and, given that it was my dream, I mean that literally!

I made the decision to go with the Asphodel of the Iliad. I cannot imagine that those three marvelous men would be anywhere except in some sort of heaven. As for the two Roswell witnesses, they are lovely people, in my estimation sacred people. (Although they would laugh at that designation!)

I don't have the impression that the other world is imaginary. Nobody's imagination can function like that for days, not producing an endless supply of detail that rich. Not to mention the fact that, when I examined something along the roadside—a flower, a tumble of stones—it was still there the next day, exactly as I remembered it. No, I think that I really was seeing into another reality, and I would suppose that others have, too, and its twilit appearance is what probably led them to conclude that it was inside Earth. But just as we have left the gods, sylphs, ghosts, fairy folk and our other interpretations of the visitors behind, perhaps it's time to rethink other folklores and legends and consider instead that some of them at least might be attempts to explain phenomena that were really observed but could not be understood. I saw another version of the world. It was oddly like this one but not quite. The observation was protracted, lasting days. There isn't anything in the literature of hallucination to explain it. So I have to classify it as an observation of an unknown phenomenon that appeared to be another reality similar to this one and apparently occupying the same space.

I have suspected for years that just such a place might be involved in the close encounter and UFO experiences—not just a distant star or galaxy in our universe but another actual, physical universe that is part of the same creation as ours is and might, thus, be connected with it. It seems possible that, if this other universe exists, that both should be thought of as a single unit, with the two halves functioning together, but in ways that we have not yet detected—or rather, only just begun to detect.

But a companion universe? Really? This would not be part of what is known as the multiverse, which is conceived of as an endlessness of universes outside of our own but not sharing the same space. There have even been a few indications that such universes may exist. In December of 2015, cosmologist Ranga Ram Chary published a paper concerning anomalies in the cosmic background radiation in which he says, "A plausible explanation is the collision of our Universe with an alternate universe." He also states, however, that "deeper observations are necessary to confirm this unusual hypothesis."

But could there be another universe immediately present, right here, sharing the same space as ours?

There is some very interesting evidence that such a thing

might be real and, not only that, a method of testing for its presence has been devised.

If another universe is entwined with ours and it is possible to cross back and forth, it would not only explain a lot about the behavior of some of our visitors and the craft they seem to be using but also reconcile some serious anomalies of physics.

Physics calls such a universe a mirror universe. And even the idea that everything in it would be the opposite of what is present here might be the case, at least to some degree. The landscape didn't look to me like the exact opposite of this one, but I cannot be sure of that. It was very similar, but I am not so sure that I would have been able to perceive it as a mirror image, even if it is one. Having no experience of such a thing, I also have no idea what it would look like.

When the two universes were created, physics tells us that the mirror must have been cooler than ours, otherwise some of its matter would have leaked across the barrier between the two, and gravity in our universe would be stronger than it is. The greater coolness of the companion universe would mean that it would have lower luminosity—just as did Homer's Asphodel, and just as I observed at Pine Ridge.

A charming hint that it may at times be possible to cross between them comes from an obscure 12th century source. It is the story of the Green Children of Woolpit. One day around the year 1130, the villagers of Woolpit discovered two children, a brother and a sister, standing beside one of the wolf pits. (These were intended to trap wolves, who abounded in Britain at the time.) The children were green in color, wore strange clothing and spoke an unknown language. They gradually adapted, lost their green color and learned English. The boy died but the girl survived and said that they came from a land where there was no sunshine and the light was like twilight. The girl was given the name Agnes and married a royal official called Richard Barre.

I've been vaguely aware of this story for years, but now the perpetual twilight described in it brought it very much to mind. Did these children come from our companion universe? If so, how? There is no record of them explaining why they happened to end up in the English countryside.

They also said that the sun never shone in their world and thought of it as being underground. But they presumably knew nothing about what might actually be causing the difference in luminosity.

This may just be an old story, completely unrelated, but there are a number of quite compelling reasons, even beyond the lower luminosity observed by me and Homer and reported by Agnes and her brother, that a companion universe might be real.

First, the Big Bang should have left more of the isotope lithium-7 in our universe. According to Alain Coc of the Centre for Nuclear Science and the Science of Matter in France, mirror neutrons coming into our universe from the other would destabilize beryllium-7, the isotope whose decay leads to lithium-7. If this is happening, it would explain why there is less lithium-7 in our universe than there should be. Additionally, we frequently measure ultra-high-energy cosmic rays coming from outside our galaxy, but they carry more energy than should be possible given the distance they are traveling. Zurab Berezhiani of the University of L'Aquila in Italy has shown that, because of the lower temperature of the mirror universe, they can travel farther without expending as much energy as they would if they remained in our universe across their whole journey. If they do oscillate between universes, that would explain their anomalous energy. In addition, and perhaps most tellingly, the most developed mirror models indicate that there must be five mirror particles for every particle in our universe. This is precisely the same ratio given by our measurements of how much dark matter must exist. It would seem possible, then, that dark matter, which we know must exist but cannot seem to find despite years of trying, might actually be this mirror universe.

So do the visitors come, then, from it? If so, then they seem to have somehow devised or evolved a means of crossing the bar, as it were, between the two.

I am reminded of Robert Louis Stevenson's poem "Land of Counterpane," where as a boy he "Watched my leaden soldiers go/ With different uniforms and drills/ Among the bedclothes and through the hills…" But I do not think that this real land of counterpane is necessarily so pleasant as he imagines. I think that our visitors are much more like the fairies in William Allingham's poem "The Fairies" for "Up the airy mountain or down the rushy glen/ We dare not go a-hunting/ For fear of little men;/ Wee folk, good folk/ Trooping all together/ Green jacket, red cap/ And white owl's feather!"

We don't see them any more in the rushy glen, not now that it's lined with condos. Indeed not, but they do come right into the condos and take us on the same sort of journeys that they always have, leaving us disoriented, confused, and as often as not with a badly deranged sense of time and space. We probably see them also in the night, slipping past overhead in their great, black triangular craft marked by lights at the three angles, generally green, red and white.

Perhaps across all our history, they have been moving between the two universes and, for all I know, coming at the same time out of the distances between the stars in the mirror universe. In that sense, they might be doubly alien, and to make matters even more complex, if creatures with the same morphology exist in our universe and have also mastered interstellar travel, then we might be dealing with alien entities from both universes at the same time, in addition to human beings from the companion universe like the men I encountered in my dream or the people who came out of the night and slipped an implant into my ear without leaving a scar, or the two whom I recently met in my living room.

Interestingly, if a so-called warp drive that opened a door between interwoven parallel universes could be created, it would involve the use of a high-energy electrical field. Just such a field was present in my garage on the morning after my implant was inserted and is a characteristic of the military implant operations as documented by Helmut and Marion Lammer in their book, *MILABS: Military Mind Control and Alien Abductions*.

And then there is the matter of metal believed to be from UFO crashes that has been found to have isotopic ratios that cannot be from this universe. Analysis of such materials was presented by Dr. Jacques Vallee and Dr. Garry Nolan in Paris in June of 2017. Their data showed that some samples taken from material gathered after a UFO apparently exploded over Ubatuba, Brazil on September 13, 1957, displayed isotopic ratios that indicated that it could not have been formed in this universe and only created artificially by the expenditure of unimaginable amounts of energy. The Ubatuba story has been called a hoax, but so have many UFO cases, and now this finding that some of the material is indeed unusual suggests that the case was real. But as Drs. Vallee and Nolan pointed out in their presentation, and Dr. Vallee reiterated in a subsequent presentation held in California in June of 2018, unusual isotopic ratios do not mean that the materials were manufactured by aliens. They only mean what they mean: at present, they are unexplained.

I think that this collection of observations, physics that demands a mirror universe and empirical evidence of material that could have come from it all add up to serious reason to consider that it might actually exist and that at least some of our visitors, and humans with strangely advanced skills, come from it.

What I can offer in conclusion is that I had a lovely and mysterious experience at Pine Ridge, and if the existence of the mirror universe is ever confirmed, then I think that I might have identified at least one origination point for our visitors.

This new phase of my life, which I would describe as a period of intensified seeing, began long before I went to Pine Ridge. I did not know it at the time, but it was on December 7, 2007, at 4:53 in the morning that I began to become aware that life can be lived in an entirely new way, and that there is, emerging from where it has long been hidden along the byways of human experience, a new vision and a new world.

# 6

# THE NIGHT DOGS

I t was windy and rainy during the early morning hours of December 7, 2007. Little did I know, not until long after the event that was about to take place, the change of life that it represented. But communication with the visitors is a multi-layered and subtle business, and I have found that it is common for years to pass before all the pieces fall into place. Therefore, in my blog entry, "A Most Complex Encounter" posted on Unknowncountry.com on December 11th of that year, I had no idea that the events from the 7th that I described had likely taken place because of the contents of the previous blog entry, posted on November 27th, "A Second Universe is Discovered" or that they represented a funda-mental change in my entire relationship with the whole visitor phenomenon.

The first time that astronomers announced that they had discovered a "tear" in the fabric of the universe was in 2007. Laura Mersini-Houghton of the University of North Carolina offered the opinion that it was the "unmistakable imprint of another universe beyond the edge of our own." A controver-sial idea in 2007, and still in 2015 when Ranga Ram Chary discussed it. In my 2007 blog, I wrote, "This morning I opened the *New Scientist* and found myself reading that a second universe is apparently out there beyond our own. I have to admit that I was, quite simply, knocked speechless. This is because, when I was talking to the Master of the Key in a hotel in Toronto in 1998, he said that there were universes beyond our own. However, at the time, I rejected his statement as obviously incorrect and changed the subject. There was, in 1998, not the slightest indication anywhere in physics or cosmology that there could be other physical universes. So when he said, 'There are more galaxies in your universe than there are stars in your galaxy, and more universes in the firmament than there are galaxies in your universe,' I listened politely and changed the subject.

Ten days after posting that blog, my world was turned upside down. It began on the evening of the 6<sup>th</sup> with what I described in my blog on December 7<sup>th</sup> as "one of the most interesting perceptual experiences I have ever had." In those days, I was still waking up in a state of fear most nights between 3 and 4, as I had been since being abducted during those hours in 1985. Now, of course, all that has changed. The fearful awakenings have been completely transformed into my beloved 3 AM meditation, which has enriched my life immeasurably. In the next chapter, I'll detail how the change came about.

On the night of the 6<sup>th</sup>, we went to bed about 11:30. While I meditate now at 11 and 3, at that time I generally did it earlier, often in the afternoon, and never in the wee hours. On that night, though, I was up again shortly after going to bed, "struck with a powerful need to meditate." I could not hear the call then, not clearly, but I was responding to it. I did the sensing exercise for about fifteen minutes then returned to bed. I then report that "at 2:17 I was writing business emails." After that, in my blog I write that I "slept fitfully." The reason was that I sensed that somebody was in the apartment. Now, when I sense this, a thrill of anticipation goes through me, and I rush to sit in meditation and get into the inner stance that opens me to communication. Then, I went prowling through the rooms, halfway ready to grab my pistol.

When I did the sensing exercise on that night, "the sensation was remarkably powerful. I felt as if I could sense much more than my physical body, as if my nerves didn't end in my skin but extended around me like a living electrical field." Well, second body does, and not only that, a surrendered inner stance can enable it to enter what in Sanskrit is called *maha-mudra*, "the clear light of the void." This is a state outside of place, an absence of being anywhere that is also being everywhere. In physics, this is called superposition, wherein a particle is in all possible states at the same time. Without knowing it, I had entered this state. Since then, I've come to it a few times and have learned more about it. When in it, reality makes a new kind of sense. All of the concerns that so weigh you down in ordinary life fall away. You come to understand what Anne means when she says, "Enlightenment is what happens when there is nothing left of us but love." I wish I could say that I was there all the time, but at least I can taste it.

I did not imagine that the state I was in had attracted any attention from the visitors, let alone that it had been induced by them as part of a lesson that I didn't even know that I was receiving.

It was surpassingly poignant, as if every sweet memory that had ever touched my mind, from a scent of autumn leaves under a boyhood tree to the day I saw my parents' car crest the hill near our house, bringing my little brother home for the first time, to the moment I looked down at a girl in an office in Manhattan and heard her say, "I'm Anne," to hundreds and hundreds of other joys, from the tiny to the great, that fill our lives, but which we too often overlook.

Then I got sleepy. Very sleepy. It was just too much, as if I'd barged into heaven before being called. I went to bed.

As soon as my head hit the pillow, a completely new gang of visitors came rollicking into my life. I wrote, "The moment I fell asleep, I had a dream that Anne had inexplicably, but out of the kindness of her heart, let a pack of feral dogs into the house…" To this day, I remember how surprised I was when what I interpreted as a pack of small, fast moving black dogs entered the bedroom and went swarming under the bed.

I leaped up and looked under it. There was nothing there. I now felt that the house was full of people. Instead of grabbing a gun, as I would have done in 1985, and did do in 1989 when I went chasing after the people who'd implanted me, I went quietly and even fairly calmly into the living room to investigate.

I was "confronted with what was just about the surprise of my life." The first things I saw as I entered the living room were three large square planters with miniature trees in them. We had nothing like that in our living room. I thought, "Holy God, I'm outside" and immediately turned around to go back into the bedroom—and found myself facing a wall. There was no bedroom anymore. Now I was scared. I feared that I'd gotten lost in some very strange way. Now that I have seen with my own eyes that there is another universe here, I'm not too surprised that I saw that wall. I was in the other version of reality, maybe even the same one that I saw at Pine Ridge. If I was really in a mirror universe, perhaps the door was now behind me.

Before I go on, I'd like to reflect that our history since the 15th century has been one of steadily discovering that the physical universe is larger and larger and larger, causing our place in it to feel ever smaller. Before that, we had many different ideas about where we were, but they were all basically Earth-centric. The moon, the sun, the stars all revolved around Earth, and we were her masters. Now we know that there are trillions of stars in trillions of galaxies, and that this is probably only one of an effective infinity of universes... which all probably have an infinity of mirror universes breathing neutrons back and forth between their realities like great, enigmatic hearts. And then there is this little band here on this tiny speck of dust, touched with intelligence and struggling to find our magic as we sail through infinity on the coattails of a wandering star.

Perhaps the next step is to discover for certain that informed speculations like the mirror universe and the multiverse both actually exist, and that reality is, in fact, infinite. If we are therefore eternal, as we must be if each of us is an infinity of selves in an ever expanding mass of universes, then that might explain why, when I asked one of the visitors back in the late eighties what the universe meant to them, I received such a startling answer.

When I used to walk in the woods at night in search of meetings with them, I had only very little success. I don't recall anything like a face-to-face sit-down, although every so often it seemed to me that there was somebody there. One night, during such a moment, I asked aloud, "What does the universe mean to you?" At once, my mind's eye was filled with the vivid, clear and deeply shocking image of a coffin.

I realized that if you knew that you were lost in infinity and could never reach either the end of reality or the end of yourself, you might well feel paradoxically claustrophobic. Something you can never leave is a trap, no matter how big it is. But, of course, we don't feel this way. We don't know that we're lost in the stars. Where they see a coffin, we see *Star Trek*. Maybe it's not as realistic a state to live in, but I much prefer it, and if they could, I think they would return to it. But when you have opened a door of knowledge in yourself, there is never any turning back.

The next thing I knew, I was lying in bed again, but not in any sort of comfortable state. Instead, I was dreaming that I was living five different lives at once. Later, I assumed that, if this had been a real event—whatever that means—they were unfolding in parallel lives, if not in parallel universes. One of them was this life, as I was living it then. I continued, "The five of them were distinct, and I was inside five different selves at once. There was no confusion, and I wasn't on the outside looking in. I was living these lives all at the same time." My most vivid recollection, looking back, is how normal this fantastic state seemed while it was unfolding.

I wrote, "In four of these lives, Anne was also present, but not in the fifth, and that was a life I very much wanted to leave. In it, I was walking down a path with a small boy, toward a quay where there were a number of men." I went on, "In this universe, Anne had died of her stroke, and I was walking with my grandson, who was about three."

At the time, I didn't have a three-year old grandson, and when my oldest grandson was three in 2010, Anne was still fine. When my youngest one will be three in 2021, she will have been on the other side for seven years. So that wasn't exactly prophetic.

In the next life, I was still living in my childhood home in San Antonio, which was now worn out and run down. Anne was bravely trying to scrub the kitchen floor. The house has long since been torn down and replaced by another house. It had been in 2007, too. In a third reality, we were living together in an apartment, and the dogs were under the bed. Fingers had grasped my hand and were tugging at it.

I now know that both the presence of dogs and the tugging on my hand had to do with things that would happen in the future, so I assume that, if these were actual parallel universes and not simply possible alternate lives, that this has to have been the one I'm actually living in. As we shall see.

The fourth reality was the one where the trees were in the living room. In this one, TV transmissions from another planet were being regularly picked up and rebroadcast by SETI, and this was a decorative motif from the other world, and I have to say that I hope that this actually happens in this universe at some point, but I hasten to add that I am not going to be buying into the idea of having trees in my living room.

At 4:53 the yanking on my hand grew strong enough to wake me up. Pulling it away, I turned over, saw that there was actually nothing there, and decided—absurdly—to try to get some sleep. The blinds were slightly open, and as I turned over, I saw lights outside the window. The wind was blowing in from the sea and the clouds were racing, but these lights were dead still in the sky and close by. I immediately woke Anne up, and we soon discovered that we could both see the lights, but only when we were in a certain position in the bed. From any other angle, they were invisible.

At the time, there was a big sensation about what long-time anomaly researcher Linda Moulton Howe was calling dragonfly drones, which were huge, complicated machines that had either been photographed over isolated areas in northern California or were elaborate hoaxes. So I thought maybe that this thing was a dragonfly drone. After a short time, the lights glided majestically off toward the ocean, moving easily against the wind and not bouncing or strug-gling in the slightest. Like the objects that had been filmed off the Nimitz in 2004, whatever was holding this object or objects up in the air was not a wing or, say, a balloon. It was not aerodynamic.

I felt as if I had been, as I put it in the blog, "moving seamlessly" between various universes. I cannot say now that I know what was happening, but there are obviously quite a few other possibilities. I could have been experiencing some sort of mind control or hypnosis, or been drugged, including with substances unknown to us. Some very strange critters were in the apartment. As I have gotten to know them, I have come to understand that they have a stunning mastery of mind and space. Since that day, I don't think that I have spent much, if any, time with any other sort of entity, not until very recently.

I don't know what they are. As far as describing them is concerned, I have only glimpsed them. I can say that they are very small, about the size of a miniature terrier. They are not dogs, but they run in packs and race around at breakneck speed. When they are near you, they are able to plunge you into all sorts of different versions of reality. But what that means—let alone how real the realities are—I have basically no idea.

As I have gotten to know these entities, I have come to feel the deepest gratitude to them. I have had three different forms as teachers: the grays, the *kobolds* and now these nameless unknowns. I am not aware of anybody else who has described a striking detail about them that I have observed, so I'm going to leave it out of this text. If anybody else has encountered them, they will know this unmissable detail. In any case, their ability to affect the mind is breathtaking. They created, or induced, my awareness of the different universes with impressive skill. They are also extremely fast. Unlike the others, they run in groups, from as few as two to many. I have never seen one of them alone.

I think that they are the ones I do the sensing exercise with now. Their intensity, full of desperation, drives me to work harder than ever before, striving to extract the sense from a life that has become so unusual that it is all but impossible to describe in the kind of practical way that seems essential to success.

Their response has been to provide me with a series of increasingly spectacular and insightful experiences, such as the Contact in the Desert and Pine Ridge experiences that now form such an important part of this book. Like this experience, they concern other universes, meaning that at this point over a period of twelve years, this has been a consistent message.

There is still another experience, described in *Breakthrough*, of driving into another world in a Jeep Cherokee with another family's little boy in the car with me. I was taking him from our country place near Woodstock to a diner on Route 17 in Paramus, New Jersey, to be picked up by his father. I took a familiar turn off the divided highway to loop back to the diner where the father was sitting in his pickup waiting for us. To my shock, I found myself on an entirely unfamiliar road. We then spent some minutes driving around in what appeared to be another world. The streets were wide and overspread by lush trees. Set back in lawns were low sandstone colored structures with deep reliefs of serpents on them. Each one had a low arched doorway blocked by a wooden door.

The boy panicked and tried to jump out of the car. He kept pushing up the automatic lock and I kept pushing it down as I drove through the broad, silent streets looking for a way back. I finally found myself driving across a sort of wasteland and ending up on Route 80 about twenty miles from where we'd left Route 17. By the time we got back to the diner, the father, who had seen us pass by, was standing in the bed of his pickup looking for us.

The boy, who I was hoping wouldn't say anything to his very skeptical dad, ran across the parking lot yelling, "Daddy, Daddy, Whitley took me on a ride through the Twilight Zone!" To make it all even stranger, if that is possible, Twilight Zone creator Rod Serling's home was just a few miles away.

We never found that eerie neighborhood again.

I wonder if it is in the same otherworld that I saw in Pine Ridge in 2019 and perhaps wandered through on that wonderfully weird night in 2007.

Looking back, though, it would seem that movement into other universes has been a consistent feature of my experience. Perhaps somebody has all along been trying to tell me something. Not only, this message goes, are you not alone, you cannot be sure where you are or even, truthfully, what you are, or what powers lie latent in that mysterious human mind of yours.

Ah, but somebody knows, and they seek to communicate that knowledge to us—not only about who and where we are, but also about who and where they are, and what it will mean for us to finally meet in what that gentle genius I was and am married to called "communion."

# 7

# THE RETURN OF THE VISITORS

Deep one night in October of 2015, pain—severe—radiated through my left second toe. Anne had passed away just a few months before, and as I had every night since, I had spent my meditation session at 11 calling to her, "Annie, Annie, if you hear me at all, please come, please come to me."

I leaped out of bed. I did not connect this with my hand being held back in 2007. In fact, it would take four more years and at least five drafts of this book for me to see the connection.

I stood gasping, then fumbled to turn on the lamp. But what was it, what just happened? There are no electrical outlets near the bed, no wiring or circuitry at all. I grabbed my phone and looked up symptoms of gout. Not a fit. I sat on the bedside, reached down and rubbed my toes.

Everything in the apartment seemed normal. I looked at the clock: 3:25. Great, now my night's sleep was ruined. Next, I looked under the bed, but there was nothing there that could have shocked me. I lifted the foot of the mattress. Nothing there either. Finally, I turned out the light and lay back down. All was quiet, the bed was warm. I drifted into a sort of half-sleep.

During the day, I thought little about the mystery. I had no idea what happened. But somebody did. They were here in 2007, and now that my situation had changed, they had returned to start a new lesson in the course of study that is my life. Without any idea that this happened, I have crossed a threshold—or rather been zapped across it.

The next night I felt strong fingers grab my right nipple, pinch it painfully and shake it. This time, I came roaring out of the bed. Once again, I fumbled for the light. I stalked through the small apartment. There was nobody here but me. All the doors and windows were locked. But that was a hand, those were fingers touching me.

You have every right to wonder, "Why is he being so dense?" The answer is that relationship with the visitors is both so improbable and so hard to grasp. They always seem to show up unexpectedly. But there's more to it than that. We may say we want to see them. We may even beg them to come. But actual contact is apocalyptic. It means tremendous, overturning change, and that is very threatening to the ego, and it is going to defend itself against what it sees as an unknown threat. This is why so many people can't take the close encounter experience. It is why we have been fighting the visitors for nearly a century on so many different levels. It's ego, defending its very existence—and all for nothing. There is no destruction of ego involved, and when you come to see that what you imagined was your "self"—the beginning and end of you—is actually just a social tool with a name attached to it, you realize that you're not really under any threat at all.

I sat down in the living room and tried to calm myself. I had finally realized that something extremely strange and yet very familiar was happening to me. There was no question in my mind but that somebody grabbed my nipple. Given the life I have lived, there could be only one explanation: the visitors were back. They had been pretty much in the background ever since Anne's fatal illness began in 2013. She had been dead now for six weeks, and I was in a state of blackest grief.

I sensed that this wasn't just an anonymous "them," though, and here began a new level of my experience. I cannot say exactly why, but I knew that Anne was involved. Since her passing, I have learned more about how deeply true her insight about the relationship between the dead and our visitors was. We are not just having a close encounter with what appear to be nonhuman beings but also with ourselves.

If the mirror universe is where what appear to be aliens come from, then maybe it is, just as Homer thought, also where our dead go. After her near-death experience in 2004, Anne felt that there was a sort of breathing between this and another universe, and that when we died here, our consciousness was transferred to another version of ourselves there.

I sat on my bedside. The feeling that Anne was there was now very strong. It was as if I could almost touch her, and how I longed to! But there's more. A dissonant note, at least, dissonant to a man enveloped in deepest grief. I sensed that she was laughing at me. In life, she always saw me as entirely too serious.

I didn't sit on the bedside for long. My mind went back to the last time I was woken up by being jostled or otherwise disturbed in the wee hours. This happened back in the 1990s when seven people who indicated that they were from between lives kept trying to get me to meditate with them at this hour. I did it for a few weeks or months, I forget how long. Then we lost the cabin and moved to Texas, and I saw no more of them.

I recalled the weeks I spent mediating with them as being

one of the best periods of contact of my life, a rich learning experience. (I've discussed it in more detail in both *Solving the Communion Enigma* and *The Super Natural*.) After I entered the meditation room the first time, they called me, they came pounding down onto the roof, making seven loud thuds. Then they fell silent. A few seconds later, I had the impression that somebody was standing right in front of me. I explained that I couldn't meditate with invisible people present. I had to see them. When nobody materialized, I left the room and went to bed.

A few hours later, one of them, a man, human appearing, became visible for a few unforgettable moments while sitting on the foot of the bed.

Their ability to control their density might involve a natural process or technology, I cannot say. If they are coming and going from the mirror universe, then maybe they don't disappear at all but simply pass back into the other reality, perhaps using some form of mental process or, of course, technology. (How, I wish I knew!)

There is one thing that, if we could do it, we could control our own density. This would involve increasing and decreasing the space between atoms—in other words, controlling the gluons that mediate that space.

The physical world is a near-vacuum. For example, the atoms that make up most of the mass of a piece of steel are actually just 0.0000000000001% of its volume. Statistically, physical matter can hardly be said to exist. It is maintained by what is known as the strong nuclear force, which is the only reason that the world we live in is here.

Their ability to rearrange atoms has to mean that they can control the strong force and, thus, may be in possession of technology that can alter density. Of course, there are other possible reasons, too. This gets back to the mirror universe, which would necessarily occupy reality in a way that mirrors our own place in it. If this is true, the math of wormholes tells us that passing back and forth might be easier than would be using a wormhole to go to another part of our own universe. The amount of energy needed to bend space-time in order to bring two points in the same universe together is far greater than the amount needed to briefly open a hole in the membrane between mirror universes.

When the man materialized before my eyes, I took his hand. It was small and light, very light. But it had definite heft. It felt solid. But I wondered if he was actually, physically present. How could he be? No matter how real it appeared, it had to be in my mind.

So I held it to my nose and smelled the back of it. Once again, I was surprised. His skin was pungent. There was a sharpness to the odor that I associate with people who don't bathe. Frankly, the guy was ripe. There's no other way to put it.

I was so surprised that I dropped it—whereupon he winked out of existence. I sat there completely flabbergasted. Now, looking back to that event of more than twenty years ago, I would think that controlling the strong nuclear force may even be a natural ability, perhaps even one that can be found, with disciplined concentration, within ourselves. If so, then it must have something to do with control of attention. It really did seem, in that moment, that my dropping his hand had broken his concentration, which is what caused him to disappear.

Fast forward to 2015 and to the third night that I was called. The first night, I experienced a shock that was inexplicable to me. On the second night came a pinch that made me realize that I was once again in contact. And now came the third night and another awakening at 3. No question now.

I got up, went into the living room, took my seat and began to move my attention from mind to body.

Since then, the early morning meditation has become part of my life, and with some of this unfolding so very close to physical reality. From October of 2015 until April of 2019, they woke me up every night by blowing on my face or the back of my hand, sometimes by kissing me.

It's easy to say, "He's just hallucinating," and ignore me. Many people don't even want to think about a life like mine, let alone entertain the idea of living with demanding invisible beings who refuse to allow you a full night's rest and who involve themselves in your inner life in ways that are often extremely challenging. But there's another, more fundamental reason they prefer to doubt me. It is that whoever or whatever is here doing this is obviously in possession of extraordinary capabilities, and that feels like a threat.

During the October–December period in 1985 in upstate New York when they were taking me physically, I felt captured, which was why I reacted like a wild animal that had been snared. And in fact, this is what I was. We are a social species, but that doesn't mean that we're not wild. We don't normally feel that wildness, but when one is face to face with an unknown being who is obviously in control of the situation and whose motives one cannot even guess at, it will come out.

This response is instinctual and is probably the underlying reason that we so generally reject this whole experience. The only way to overcome it is for both sides to keep trying to get used to one another.

There is a taming process necessary, and it isn't easy, not even when you know what is happening. I've been trying to get used to them for thirty years and have only just recently begun to think of my fear of them as something that has passed. Frankly, my wife's participation in the process from the other side has more than anything else been what has enabled me to make progress, and, to me, this is an indication that contact isn't going to succeed if we continue to deny the existence of our souls and thus also our dead, and they are not involved. We need to finally stop pretending to ourselves that they don't exist and get down to the business of enlisting their support as we go deeper into this new life.

Here are two examples of just how deep and powerful our fear of doing this is. In February of 2017, I was at the Esalen Institute in Northern California at a conference with Jeff Kripal. We were sleeping in the same room, one in which I had briefly encountered the visitors on a previous visit. It's in a building called the Murphy House and is called the Sea View Room because it has a deck that overlooks the Pacific Ocean. When they are going to drop into a physical density, the visitors do tend to prefer spaces that offer a quick exit to an open area such as a large forest or the sea.

At 3 on the last morning of the conference, an invisible presence blew on the back of my left hand. I was lying with my head turned to the right, which meant that I was facing toward the window that overlooks the sea. As the burst of air on my left hand caused me to open my eyes and turn toward the hand, I glimpsed a dark figure at the bedside on my right. I saw no detail except that it was short, and I assume that the fact that the left hand was involved was intended to cause me to turn away from it as I awoke. Even as recently as 2017, I would have reacted to face-to-face contact with a burst of fear.

The next moment, I recovered myself. I got up and, as usual, opened my inner self with the sensing exercise. Nothing further happened to me, and I completed the exercise, as I normally did in those days, after about fifteen minutes of inner work.

The next morning, Jeff remembered hearing a tremendous

crashing sound and feeling an uncanny sense of dislocation. My experience had come around 3, and his about an hour later. He heard an inner voice that was at the same time his own say, "Oh my God." He told me later that he felt a sense of devastation, as if his entire world was collapsing.

This is because contact involves the breakdown of the barrier between the living and the dead and at least a partial drawing of the ego out of the time stream—a sort of death before dying. This threatens a fate that is horrifying in the extreme to the inner person, which is being plunged into the nonmeaning that accompanies knowledge of future and past.

The natural reaction is to think, "Oh, I'd love to know my future."

But what would that actually be like? In fact, you'd feel like you were riding on rails, or a marionette being manipulated by an unseen puppeteer. Your spontaneity would be lost. Life would entirely lose its meaning.

I think that this is why Jeff was so devastated in his moment of contact and why I and so many others have struggled with the ferocious, nameless fear that comes to us when the visitors approach.

The fear can be much worse than what Jeff experienced. It can be life threatening.

The previous summer, I had been at a country house where I've been encountering the visitors since boyhood. It has a sleeping porch upstairs, which a number of the bedrooms open onto. An individual who was in one of these rooms heard scraping footsteps outside her window and then a low, husky voice growl, "Why aren't you asleep?"

She called out and asked if it was me. I was in the living room and called back that I was downstairs reading. I had also heard those footsteps, though. When I was a boy, I heard them on that same porch many times.

In the morning, I asked her if she had been upset. She said no, but I knew from experience that an encounter like that, even one so small, can have powerful effects.

Sure enough, that afternoon she began to experience symptoms of what I suspected could be a silent heart attack. We called a doctor friend who lives nearby. He came over immediately, confirmed that a heart attack was in progress, called EMS and got her to the hospital. She ended up with a pacemaker.

These are typical examples of the kind of stress close encounter produces. I was allowed on the night at Esalen only a glimpse of what was there. Any more than that, and my ego would have felt itself being drawn out of time, and the terror would have come.

Understand that this doesn't just happen when we have contact with nonhuman intelligences. The literature of ghostly encounters with dead human beings is a literature of fear. But although both types of event are among the most challenging experiences a person can have, both can also be not only endurable, but also productive.

We can see a version of our own fear in the fear that wild animals have of us. In the distant past, I would think that they were no more wary of us than of other predators, and larger predators weren't afraid of us at all. That was, however, before we came to understand the inevitability of death. Because we know this, we are now different from all the other creatures on the earth, and they know it because they can see it as a darkness in our eyes, exactly as we see in their terrible glances the visitors' knowledge of the future, not just that death is inevitable but the day and the hour.

The visitor who caused me to turn away from him did so out of kindness. If I had woken up while I was turned toward him, I would be looking right into his face. A moment or so of that, and all spontaneity would be leached from my life, for knowledge of the hour of death must shed a cruel light along the path of future life as well.

I knew a man who had something close to this happen. After looking into the eyes of a visitor, he spent the rest of his life in a state of permanent *déjà vu*.

Once you know the moment of your death, you also know everything that transpires between. We are not here to move through life on the grim rails of future knowledge but to experience events spontaneously. Even if they are preplanned, which for all I know they may be, our purpose here is to be surprised and to gain self-knowledge by observing the way we react to what life presents to us.

If the visitors, not to mention our own dead, are going to commune with us, they are going to have to hide very carefully, because if they slip up and cause us to lose the chance to react spontaneously, then they also lose what they are here to experience.

This is the primary, but deeply hidden, reason for all the secrecy that surrounds the contact experience. Contact—communion—involves not only a new kind of intimacy but also a new kind of mutual discipline. We have to open ourselves to them without seeking to know them, and they have to enter us without revealing their knowledge of our futures to us.

Without us having confidence that this won't happen, communion can only go so far, and it's not going to be far enough for either side.

The question remains, "If it's all predestined, why bother to do anything?" The answer has to do with the reason that history itself exists. Over the six or seven thousand years of the current cycle, in each generation, more complex lives have become possible. The number of alternative histories on offer to each soul has become larger and larger. Even though the end is always the same and the game remains the game, there are more possible moves in every generation, more twists and turns. Although the end is always the same, the paths of life become more and more forked, and the journey richer in discoveries.

Knowing all that, though, what practical means do we have to work with the fear?

Just as tame animals don't fear us, we can learn not to fear the visitors. That's a great part of what this book is about. In February of 2017 at Esalen, I dared not look into their eyes. Now, in Santa Monica in August of 2019, I understand why I shouldn't and, therefore, have no fear of this. I'm not going to ruin the spontaneity of my life when I'm with them because I know how to avoid doing it.

I get knowledge from them. My life becomes richer. They get relief from me. Yes, it's a game, and just as Shakespeare said, this is a theater and we are the players. He didn't think about the audience, though. They are the audience, and when they have a seat in the theater of our lives, they enjoy the great pleasure of being alive again.

It's an illusion, of course, but it must be a satisfying one because they do come back to me for it very often indeed, and they are clamoring for more opportunities with more people. If we can do a good job for them on the little stage that is this earth, they are going to help us keep the theater open—that is, help us rebalance nature.

It's going to be hard to make the relationship work, though. Probably a goodly number of my readers are recoiling in horror right now, thinking to themselves, "My god, he lets himself be possessed!"

No, shared. If they controlled me, they would bring the knowledge they want to leave behind with them. A lot of close encounter witnesses intuit this. It's why they are so often called "The Watchers."

Communion is a new state for us. From experience, once one relaxes into it, the idea of living in the old way becomes the real terror. This is especially true because of the alternative, which is upheaval, incredible human suffering on an unimaginable scale, and possibly even extinction.

**8**
___

# THE MAN FROM PARADISE

Until recently, I did not understand the multiple witness encounters that took place at our cabin for what they were. It is only as I have gotten into deeper contact with the visitors that I have been able to understand that they were messages and decode them.

As I've done this, a whole new picture of who the visitors are and what they want has emerged.

The first of the big cabin events involved the *kobolds*, the second the grays. On the first weekend, there were ten adults present and one child. Anne, Dora Ruffner and Peter Frohe have since passed away. Ed Conroy, who was then writing *Report on Communion,* Lorie Barnes and Raven Dana have agreed that I may use their names in this book.

Raven has been kind enough to send me an email detailing her memories of what happened on that night in the living room where she, Dora, Peter and another man were sleeping.

Lorie was in a guest room along the hall, our son in his bedroom, and Ed and his companion in the basement. Anne and I were upstairs in our room.

Raven writes, "Dora and B. were maybe five feet apart,

sleeping parallel to each other. I was sleeping with my back to them. I woke up because I had become hot and uncomfortable. When I tried to roll over, I found that I couldn't move. I was wide awake though and thought, 'sonofabitch!' I tried to relax. Then I heard Dora and B talking but I couldn't make out what was being said over the driving rain." (A sound like rain would often fill the house as the visitors approached from above.) "Then I heard some thumping, bumping sounds. The rain let up and I heard birds."

"This all happened over a period of just a few minutes. Heat. Heavy rain. Could not move. Muted voices. Thumping. Then silence and I could move. My eyes were open the whole time. I rolled over and said, "What the heck is going on?""

"Dora and B. both started talking at the same time… talking over each other…me, too. I asked, "Did you hear that rain? What was all that noise? What time is it?""

"B. said that the visitors were just there, and had been doing 'acrobatics; around the beds. Dora said it was 2:40 and that she didn't think what we heard was rain. We all went to the door and…nope...there had been no rain. Everything was dry and we could no longer hear birds."

"Apparently when I could not turn over, several small blue beings were jumping on or around B.'s bed. That is the noise I heard."

While this was taking place in the living room, in the basement, Ed and his friend were astonished to see a young woman walk down the stairs holding a badly torn up sweater. It was a close friend of theirs, who had died in the 1983 Mexico City earthquake. She had been so terribly crushed that all that was recovered of her had been the sweater. Here she was, though, seemingly perfectly healthy, carrying it! (This is very typical of contacts with those in the afterlife, by the way. Most of us cannot hear them, so they will come carrying objects that identify them.) After conveying the information that she was all right, she disappeared. Upstairs, the fun went on for about fifteen minutes more, then the dark blue acrobats were gone.

Pretty much everybody in the house stayed up talking all night except me and Anne and our son. We were in our bedrooms and slept through the whole thing.

As I now understand, it was a communication on two levels.

It was the first time people other than ourselves and some of our son's playmates had witnessed the visitors at the cabin. It was also another indication of a connection between these entities and the human dead. I say "another" because, during the *Communion* experience, as I said in *Super Natural* and elsewhere, a dead friend was present during the initial phase of my abduction. I didn't put it in the book because I didn't realize its significance. I still didn't, and that would not change until Anne pointed it out.

About a year later, filmmaker Drew Cummings was there making a documentary about the *Communion* movie. Raven Dana and Lorie Barnes were once again present, A third woman was also in the group that night, as were Ed Conroy and Dora Ruffner. Drew had brought a low-light video camera and planned to set it up in the house, so we were very hopeful that an event would take place.

Late in the afternoon, Lorie came in from a walk looking bemused. She said that she had just encountered her brother on the road. What made this so amazing to her was that he had been missing for twenty years and declared dead by the FBI. He had appeared in a brown robe and hood, much like some of the visitors do, standing in the woods just beside the road. Lorie asked him to come down to the cabin to meet her friends, but he said only that she was in the right place, then drifted back among the trees and was gone.

Knowing by this time that the appearance of the dead was

often associated with the coming of the visitors, Anne and I began to think that the visitors might show up later that night. We said nothing to the others about this, of course.

We knew by then that dancing and chanting would sometimes bring them, so we went out to a cave where I often meditated in those days. It was about a mile from the cabin, in a cliff above a little stream. It was a challenge to get to it, and once you were inside, you could not be heard if you cried out and you could not easily leave.

Except for Cummings and his crew, we all went to the cave, where we chanted in a way that I had been using for some years. This is called overtone chanting and requires a powerful use of attention, concentrating and letting go at the same time, in such a way that the vocal chords vibrate differently and the voice can produce harmonics. It's a Tibetan Buddhist discipline and I had found that concentration like this could sometimes get the attention of the visitors.

When we got back, we talked for a while, then Anne went upstairs. Lorie and the other woman went into one downstairs bedroom and Raven into the other. The Cummingses bedded down on the convertible couch in the living room so that he could attend to the camera, which was to be left running all night. I camped out in the woods with our son because Lorie had his room. Ed and Dora did the same.

Sometime later, Raven was awakened by movement. The first thing she saw was an Eye of Horus on the wall. It was not a hanging. It had not been there earlier. Then she noticed that what she at first took to be a raccoon had come in the window. An instant later, she remembered that the window screens were all screwed closed, so it couldn't be anything normal. She realized that it was one of the visitors. When she did, it reached out its hand, and they touched. This was a rare incident of physical contact with a person in normal consciousness, and it sent delicious yet powerful energy pulsing through her. She heard it ask, in her head, what it could do for her. She replied, "You could go down that hall." (Where the low light camera was now in operation.) It then disappeared.

A moment later, Lorie was woken up by being punched on the shoulder. She saw the entity staring down at her, but a moment later it was gone.

In the living room, Drew then woke up to find a small man with a large head peering down at him from beside the bed. He was startled, of course, and when he reacted with surprise and fear, the man's head turned into that of a falcon. Then it disappeared.

Superficially, this would seem to have been nothing more than another bizarre event in a cabin that was at that time really a sort of haunted funhouse.

Let's look a little deeper.

The first sentence had been uttered about a year before this when a carpenter who was working on a house up the road from our place decided to spend the night there because the house wasn't yet sealed and all of his tools were inside. In the middle of the night, he woke up to movement and was appalled to see a strange little man, dark in color from head to foot, standing across the room staring at him. As he jumped up, the man changed into a bird of paradise and then disappeared before his eyes.

In the next chapter we are going to dip into ancient texts, specifically into the Pyramid Text in the Pyramid of Unas, to learn more about visitor communications, like this one, that are pictographic and representational. Because we no longer use pictographic languages, it's particularly difficult for us to pick up on this aspect of the way our visitors express themselves.

Before seeking to uncover the meaning here, let me offer a thought about why hieroglyphs would be used at all. In this situation, the hoary old adage "a picture is worth a thousand words" could not be more appropriate. Looked at as imagistic communication, these few brief appearances, each one lasting only a few seconds, left behind a treasure-trove of information. This quality of compression is a consistent characteristic of visitor communications. Even when words are used, multiple meanings are conveyed. For example, when the words "A new world, if you can take it," were said to Col. Philip Corso, both meanings were important—if we can wrest it out of their hands and if we can bear what we find.

First, we're shown a bird of paradise, then the falcon god Horus. The connective tissue is that both visitors had the same general appearance and both transformations involved birds. The first sentence, the transformation into a bird of paradise, is straightforward: "I can fly like a bird, and I belong to paradise." The next one, Horus, is more complex. First, the term in Egyptian mythology represents a number of avian deities, primarily Horus the Elder and Horus the Younger, two different gods with different attributes. But there were also many more granular manifestations of the deity. This suggests that we should think of the entity as being part of a larger group, perhaps an entire civilization, a whole species, a world. The pharaoh, during his lifetime, was identified with Horus, meaning that the entity belongs to life and to what we might think of as kingship or leadership, and a noble tradition. The falcon hieroglyph refers to the star Sirius. As well, the falcon is the fastest animal on Earth. It also ascends the sky in circles, just as energy does when it rises up the spine. The falcon is also the ascending soul.

So we have in these two brief images what amounts to a self-portrait. A remarkable being is telling us about himself in a language of transformations and images from our own ancient memory and from nature. His first sentence, spoken to the carpenter, told us where he was from and how he could navigate. His second is "I am a king and also very fast." Then, "I am from Sirius, and I am returning." Then, "I am a living soul." All of that said without a word spoken or written down, but eloquently clear if one accepts that there could be a language based on signs and imagery that work like hieroglyphics and that somebody could make the words not by writing but by changing their own appearance.

This is another example of why, in exploring life with the visitors, it is so important to step out of our accustomed ways. Maybe we will talk to them one day on our own terms. But we can never have a conversation with them—or, for that matter, among ourselves—like the one described here because linear language simply isn't rich enough to communicate on that level. If we are attentive to this method of conveying information, we can learn a great deal from even brief exchanges.

But who was he, really, and is this paradise actually in another star system? Could it be that simple?

Actually, it might be even simpler. He was announcing himself as part of enlightenment, a universal possibility that is shared by every one of us. In this sense, he was expressing continuity with us, because we are on parallel journeys. We both seek paradise but not on a distant star. We seek it here and now where the kingdom of heaven, or enlightenment, lies, as some religious texts assert, within us.

Looking at the world around us, full of hatreds and violence and in the process of failing in fundamental ways, it seems impossible that such a path could even exist, let alone that we could travel on it, let alone that we could ever find the kingdom, the paradise within to which his example may seek to guide us.

But if it isn't possible, why demonstrate it? I doubt that he would even be here if it wasn't. What would be the point? This is why I consider these rough, frightening beings to be midwives to a new birth of mankind and a new world. On the way down the birth canal, both mother and baby struggle mightily. Baby experiences terror and pain. Then he is laid on mother's breast and begins to feed in a new way. Everything that has previously been received through the umbilical cord must now be taken in through the mouth, and all of life follows. "A new world, if you can take it." Baby experiences for the first time the flavors, sensations and comfort of the ordinary world, and mother and child bond in a new way. Mother is no longer an abstraction but a person with a voice that baby loves to hear and a face that amazes, and baby begins to grow up and does grow up, and mother grows also into the fullness of her womanhood and then fades, as baby will, too, in the great river of time and the flow of life.

So it is with mother and child, and so it will be with us and Earth, whom our midwives are trying to help us discover in a new way. It is when we are born and she is exhausted from her effort that our true relationship will begin…unless, of course, we are born dead.

That is why they are here to make sure that doesn't happen. As part of that effort, this book is being written. But one little book is only a small part of such a large task.

Once they are finished with their work, we will all say, just as do the subjects of the best emperor in the eighth verse of the Tao Te Ching, "It happened to us naturally." We also, free at last from the helplessness of life in the womb, will have gained the right to climb to the star of the man from paradise, and our own truth.

But does that mean he's from Sirius? Literally?

We have explored the probability that our planet has a companion in a mirror universe, but how likely is it that there might also be aliens here from other planets in this universe?

The arguments against this have been many. The first is that nobody is likely to be able to cross the unimaginable distances between stars. The second is that, even if they have that theoretical capability, the practical benefits would be so small that nobody would go to the expense. The first I call the "lack of vision" argument, the second the "lack of imagination."

The evidence that the visitors are from other planets is not as strong as the evidence that they have some more enigmatic origin. Given all the testimony about them walking through walls, appearing and disappearing, levitating and so forth, at least some of them are functional on many different levels of reality and enter and leave the world as easily as we might a swimming pool. Maybe this is all just technological legerdemain, but it could also be that they are passing back and forth between this and the other universe.

In the predawn after the experiences of Raven, Lorie and the Cummingses, I was walking up to the cabin with my son when we saw a translucent, hooded figure come out the front door, race down the deck, across the back yard and into the woods. As it disappeared, it flashed back and forth among the trees, carefully avoiding them. At the same time, the Cummingses experienced a burst of heat so intense that, when we entered the house, they were both on their feet. They thought that the bed had caught fire. Whatever happened, these things must be true: The being had solidity in this world or would not have needed to dodge the trees. It must also have been doing something to bend light around itself while in the house, thus, rendering it invisible. We know this because of the release of heat, which would have been retained while gravity was being controlled in its immediate area, which would have been necessary to bend the light. We also know, from observing the behavior of the devices recorded by the Nimitz pilots, that devices that can control gravity are in use by somebody.

Is this somebody from Earth, from the mirror universe, or

from another world…or, once again, is it a combination of these?

The main argument against the alien hypothesis has always been the distance issue, but that has recently been called into question. A paper submitted by Jonathan Carroll-Nellenback and colleagues to the Astrophysical Journal in February of 2019 suggests that "the Milky Way can readily be 'filled-in' with settled stellar systems under conservative assumptions about interstellar spacecraft velocities and launch rates." While they assume that there are no interstellar visitors on Earth at the present time, they show how adding the effect of stellar motion to speed calculations would enable the spread of life possible on time scales much shorter than previously assumed. In the past, when things like the Drake Equation, which measures the likelihood of somebody from another planet finding us, were conceived, the fact that stars move was not taken into consideration. The Nellenback paper corrects that misapprehension, and shows that crossing interstellar distances, while still very slow by human standards, is probably far from impossible.

Given all this, it is time to stop being so certain that somebody from this universe cannot be here—and, in fact, both from this and the mirror universe, given that there is evidence for that, also, as a point of origin.

This gets me back to Sirius. According to Susan Brind Morrow in her book on hieroglyphics, *The Dawning Moon of the Mind,* the falcon glyph is associated with both Horus and Sirius. As the brightest star in the sky, Sirius is also among the most frequently mentioned in mythology. Among the Greeks, it was known as the Dog Star because of its position in the constellation Canis Major, the Great Dog. This same designation is repeated in many cultures around the world that have no obvious connection to Greece. In Chinese and Japanese myth, it's known as the Wolf Star, and among many different Native American tribes, it is designated also as the Dog Star.

Among the Dogon people of Mali, who are believed to have had some connection to ancient Egypt, there is a story that entities called Nommos that lived in water came to Earth from Sirius, bringing knowledge, and Horus, while the opposite of a water deity, is also thought of as a bringer of knowledge. Additionally, one of the forms of the Apkallus, mythological knowledge-bringers of the Sumerians and other related cultures, is that of a man wearing a cloak made of the skin of a fish.

In my own recent experience, the dog has played a powerful symbolic role. I perceived the entities who appeared in the apartment in 2007 as dogs. In September of 2019, I had a long interaction with one that appeared as a black dog. This was not a physical experience, but it was powerful. I felt as if I was being watched by a careful and penetrating mind—hardly that of what we would call a dog. I was being supervised.

I have already been warned that time is short, and the book must be gotten out so quickly that I can't seek a publisher. There is no time. I must self-publish instead and as soon as I can. The warnings have been truly fierce, but not threatening. Urgent.

So the man from paradise, it would seem, opened a door with his brief visit to a vast amount of myth and one of the great mysteries of the past, as well as, in the end, to my own experience. Sirius is so important in so many cultures and so often associated with the bringing of knowledge and ascension to the stars, and of all the visitors I have met, the ones who use the symbol of the dog to address me have brought the most knowledge.

I know that this all sounds very strange, but in communicating with the visitors, it is essential to be prepared for asymmetric methods, especially the use of imagery in ways that are not common among us. This also goes for sound.

Another example of this is the incident of the nine knocks that I describe in *Transformation*. I was sitting in the living room at our cabin when there came, in three groups of three, nine distinct knocks on the roof. These sounds were, like the three cries that I would hear above the woods, absolutely startling in their perfection. In fact, until you have heard something like that, I don't think it's possible to fully understand what it's like. This is because it's not something we ever hear. When you do hear it, though, you know immediately that you are having a new experience. My cats certainly did. Their fur puffed up, and they began creeping along the back of the couch yowling.

I cannot know if this was intended, but the knocks reflected a tradition in Masonry, where when someone is elevated to the 33rd Degree, they knock in this way on the door of the hall before being admitted. Gurdjieff's law of three and the riddle of the Sphinx are also referred to, as is the alchemical notion of salt and sulfur being balanced by mercury. The principle is expressed in the riddle of the Sphinx as courage (the lion) and strength (the bull) being brought into balance by the mind (the human head).

When the three are in harmony, then the Sphinx, as did the man from paradise, spreads its wings and soars aloft, looking down on the cares of the world with balanced and objective vision.

Close encounter witnesses, when they look, are likely to find similar communications in their own lives—richly visual, referential to the life of the soul and the increase of consciousness, suggesting that travel along our ancient spiritual paths remains a journey worth taking.

## 9

## SHARED LIVES

Throughout this book, I have been talking about a number of different forms of nonhuman and/or nonphysical beings. There are visitors who appear at times physically and at times nonphysically and which do not resemble us when they are in a physical form. There are the human dead, one of whom in my experience once briefly generated a physical version of himself. Then there is the conscious field, which I have also called the "presence" and which seems to me to be the great ground of reality, which Anne once described as the "yearning" that underlies all that is. It is this presence that generated the incidents of light that I sought in the past to identify as God and that I now see as a sort of field of conscious energy that is at the basis of all that is.

This and the next chapter are, very specifically, about our nonhuman visitors. The two chapters concern what they want and what they have to give, and what they will take if we abandon it.

Their needs are going to seem as strange to us as our lives do to them. Because of the way they are structured, as I have discussed, they cannot experience surprise. Their lives, trapped in what amounts to an eternal present, are absent all the excitement, wonder and beauty, all the pain and terror, all of the *living* that defines the human experience.

They are here for surprise, for beauty, for excitement. They, living and yet dead, are here for life. Our blindness to the future, which to them is our most precious asset, is what enables us to have novel experiences and learn from it. It is what enables us to journey toward ascension and ecstasy.

They have all the knowledge there is, but they cannot make this journey. This is why they are so desperate that we survive, for we are their main chance to participate in the wonder of life and touch its joy. They want us to join together, to cooperate with one another. We will provide the excitement of the journey, they the knowledge that keeps it from going off the rails—which, right now, it is in the process of doing.

I don't know why they are as they are or have the needs that they do. I think it's damned lucky for us, though, because without their help, I don't think that we're likely to survive. With it, we are going to become a huge, extraordinary engine of experience, travelers in every place that the new is to be found. We will cross the reaches between the stars and find new paths in the mirror universe and other universes, carrying with us our brilliant companions, who will be providing all the knowledge we need to make the human journey grand and to make it last, and to give them what they so desperately need, which is a share in our wonder.

Perhaps it's because they have learned so much that they are ranging the firmament in search of a new experience, as Kuiper and Morris speculated. Perhaps it's because they are conscious machines and cannot make the journey through heaven's gate except hand in hand with us natural beings.

However they have ended up as they are, I am rather sure that I am talking here about the core reason they want contact, and the deepest meaning of communion. Their chance to join the expanding wave front of ecstasy that is the true goal of life depends on gaining some sort of partnership with us, who cannot help but be on that journey.

As I have come to know them, I have come to feel compassion for their need. I have also found that we have common ground, and I have used that to find also a basis for relationship.

Frankly, I'm excited about the partnership. To many, it will at first be frightening. They will fear that it is possession. Of course, that's exactly what it mustn't be. They must be silent sharers, or they will not get what they want. Not only will there be no possession, they will do everything in their power *not* to expose us to the totality of their knowledge. That would ruin communion for them. Their adventure, and their joy, depends on our not knowing everything. In this sense, we're a perfect match for each other: we need their knowledge; they need our innocence.

We have the potential to share because, deep down, we are woven from the same fabric. We are both among nature's predators. As I will explain in the next chapter, I also have reason to believe that, while their involvement with us can be made to benefit both sides, there are situations where their predatory instincts prevail, and I know why.

Normally, they live among us as they do with me, in a symbiosis that is, for the most part, secret and therefore limited in ways that are very frustrating for them. Without openly sharing the journey, they must ride in silence, never taking the rudder of life, not even when it would be in our best interest that they do this.

Right now is a good example of such a time. If they were sharing their knowledge more directly, we wouldn't be on the point of extinction.

Normally, in my experience, they share, they do not take

—by which I mean, take the richness of experience that we gather into ourselves as we journey through life. From what I have seen and learned, they will remove from our souls only what we have abandoned of ourselves. This is why being awake to our lives in a richer way—the sort of acute consciousness taught by G.I. Gurdjieff and others—is so important. If you at once live your life and see yourself living it, they must share your experience. They cannot take even the smallest bit of life that enters you. But there is evidence that they feel free to take what of ourselves we abandon, and that, if we abandon ourselves entirely, they feel that they may take everything.

When you are physically face to face with them—which I have been on a few rare occasions—you feel not only the devastating power of their reflective eye, but also the sense of the predator. They're wary, too, I suspect because they sense our predatory nature as well. The difference is that they hunger for our souls while we think in terms of lashing out at their bodies.

When I first started trying to engage with them, I was just amazed at the level of fear I was experiencing. When I saw how my cats reacted on the night of the nine knocks, I was shocked. Those animals were more frightened than I thought they could be. Cats' tails puff up when they're scared, but when they're *really* scared, as they were that night, they apparently become head to tail fuzzballs. Their yowling was just unearthly. I think that they were like this because they have souls, too, and their souls were under threat and they knew it, and that is much more terrible than the threat of death.

While I do not think that we necessarily need to fear them, we do need to be aware of the nature of the threat that is present. The reason that I don't think that fear is necessarily an appropriate response comes from my own life experience: I have had them in my life for thirty years and I'm still here, still free, and living a richly satisfying life.

Another is how our relationship has evolved. In the early days, I would walk out into those woods at night almost unable to put one foot in front of the other. During the ten years I did this, I was afraid the entire time. In other words, proximity didn't help. This is because this is not about somehow getting used to them. It is about understanding yourself as well as they understand you, so that when their gaze penetrates to your darkest, most hidden places, you will not be shocked by what you see there, and therefore you will not be terrified. If my experience is any example, what you will be, once you really get to know yourself, is understanding. Your shame, your dread of your failures, your imperfections—all of that stuff that you don't want to face—will gather together in a great flood of acceptance, and you will be free. Still the same but free. And then, deep in your life and deep in yourself, you will start trying to repair what you can of any hurt you have brought into the world.

It's called being a seeker, and when you join that motley crew, you will find that the visitors, awful though they sometimes seem, are your companions.

In February of 2019, I spent the night alone in a house with them that was miles from any help and from which there was no quick escape. We had a meeting, and it wasn't pleasant. They were angry because this book was going too slowly. The meeting was not a sit-down discussion. Hardly that. But it was certainly a meeting. What happened was that I proposed to work on this book during the day and on a novel I've been aching to write at night. The answer was a decisive "no."

Twenty years ago, I would have run out of that house, gotten in my car, and driven straight into the depths of the nearest city. On that night, once our business together was concluded, I went upstairs and went to bed. I had a good night's sleep, interrupted as usual by the 3 AM meditation, during which they were so close to me that it was like being stared at by a hungry tiger from three feet away. But there was love, too. Lots of it. Yes, I was face to face with a tiger, but I am not the tiger's prey and my soul knows it.

So how did I manage this little act of legerdemain? I'm still full of imperfections. The difference is that now I'm not in denial about them. I know them and accept them, so when the visitors see them so do I, and I'm neither repelled nor surprised. I have accepted myself, warts and all.

The result is that I am comfortable sharing myself with them.

Nevertheless, it feels incredibly dangerous to let them in. I know this for certain because I have been doing it, and struggling with the fear involved, for years.

What I have come to is that I don't think they're dangerous to me, and I don't think they will be after I die, either. Anne wasn't perfect, and she ascended. I saw her do it. In fact, the great majority of us die into a higher state, all of the richness and complexity and beauty of the lives we've lived adding to the ecstasy that is the greater aim of all life.

Ecstasy, though, does not involve only pleasant experiences. It is the process of accepting all experience. Ecstasy is everything, reconciled.

It's said for good reason that the devil is a tempter, and I have had them draw me toward all sorts of angers, lusts, and so forth, simply because it was exciting. In the real world, though, that same entity, who as a demon, tempted me in ways that would have left me with regrets that would impede my eventual ascension, would also glow with excitement when I was feeling love.

They have taught me over all these years, with endless determination and patience, how to be the best man I can be, so that I can be their partner instead of their victim. It is my impression that I've lost the deep fear of them because my instincts are now telling me that I'm not going to be preyed upon. I'm no longer a potential food source but a partner in the journey. I share my life and they share their knowledge.

It is this trade, incidentally, that I think has the potential to save us. What if there were ten thousand scientists like Ed Belbruno, all receiving knowledge like he did? We're looking at another enormous increase in human knowledge, which will bring with it also the collapse of the barrier between the physical and nonphysical sides of our species, and a subsequent new vision of what death means and how to live a moral life.

Just as they are predators in the realm of the nonphysical, we are in the physical world. We won't prey on another creature if we live in symbiosis with it. Dogs and cats are not eaten because they work for us. The cats came into the granaries of the Egyptians and ate the mice and rats. The Egyptians were so pleased that they didn't just treat them with respect, they considered them gods. Dogs hunted with us, of course. With horses, it's the same. For ten thousand years, they carried our burdens. Other things that we keep, like monkeys, birds and so forth are safe from our larders because we enjoy their presence in our lives.

Nature is full of symbiotic relationships, and there is every indication that this is exactly what is trying to happen between us and the visitors. There is a difference between this symbiosis, though, and, say, ours with the dogs. It is that we, also, are highly intelligent. For this reason, there is potential here for a truly electrifying partnership, with both sides gaining tremendously from it.

For us, it means empowerment in the face of looming catastrophe. For them, it means freedom from the death in life that is always knowing for certain what your next step is going to be.

The question then becomes, "How can I become a symbiote instead of supper?"

The answer could not be more simple: become a strong soul, and it doesn't matter whether you believe in the soul or not, you can still do this.

The world is filled with texts about how the soul works, why it is there and what its fate may be. All of these texts address the matter in the context of one set of religious beliefs or another. The modern secular script, which I followed for most of my adult life, says that there is no soul. For visitors who are just looking for a nice dip into the thrill of life, that's a lovely turn of events because it leaves the victim of that belief so vulnerable.

As we explore the reality of soul life and soul vulnerability more objectively, the question arises as to whether or not there is any way to build a strong soul outside of the religious context. Is it possible to be a modern, secular person and still be attentive to one's soul?

Indeed it is, and you needn't even address the question of whether or not you have one. Living a good life as insurance will work just as well as living it with the belief that the soul is real. On the other hand, if religion is your preferred way, it does offer paths to build a strong soul that are effective. It also offers pitfalls, of course, the chief one being that the faithful should kill off everybody who doesn't believe as they do. As Anne says, "The human species is too young to have beliefs. What we need are good questions." This is particularly true when it comes to religion. We don't yet have a ground of certainty to support any religious belief that exists now or ever has.

This doesn't mean that they're wrong, and certainly not that they're useless. The most sublime text on meditation ever written, the ancient Chinese *Secret of the Golden Flower* is a Taoist religious text. It is also a brilliantly insightful exploration of how to use soul energy and increase soul strength.

When the visitors first started showing me that the soul could be understood outside of a religious context, I began looking for some way of doing this. Could there be an objective science of the soul? Was there one somewhere, or had there ever been?

I think that there is a very old text that is not entirely religious but rather addresses the life of the soul through an objective lens, which I am calling soul science. This is because it is about the life of the soul, the health and feeding of the soul and the journey of ascension, but with few religious references. It is, in other words, a sort of craft book of the soul based on what I think of as a lost, objective science of the soul.

I suspect that there was a time when we could perceive our souls more directly and had not caused them to disappear into the illusion called the supernatural. By science here, I mean the systematic study through observation and experiment, in this case of the soul, carried out just as if it was part of the physical world like any other natural phenomenon—which I believe that it is.

The text I am referring to is the 3,200-year-old Pyramid Text found in the Pyramid of Unas, which I mentioned in the last chapter. Before I begin describing how, exactly, it describes the soul and its connection to the body, I would like to discuss the energy that is involved here.

We now commonly call it things like *prana, chi,* and *kundalini*. From the fact that it can be demonstrated to work in acupuncture, some western scientists now believe that it exists, but it has never been successfully detected. I think that it goes undetected for the same reason that the visitors do: it is not going to submit to detection by anybody who does not understand that it is conscious. There is little written about it as such, but the enigmatic Master of the Key spoke extensively about it during our 1998 meeting. He said, "Conscious energy is not like unconscious energy, the servant of those who understand its laws. To gain access to the powers of conscious energy, you must evolve a relationship with it. Learn its needs, learn to fulfill them." I then asked how to do this. He replied, "By first realizing that you are not cut off. There is no supernatural. There is only the natural world, and you have access to all of it. Souls are part of nature." He also said that it was part of the electromagnetic spectrum and detectable as such, but also that it isn't passive and will decide whether or not it is to be detected, and the degree.

The visitors are full of this conscious energy. When one of them touches you, you feel waves of it coursing through your body. This can be pleasant, as it was for Raven Dana. It can be so powerful that it is incapacitating. I've experienced this. It can be painful, which is what happened when it shocked me awake in September of 2015. Coming into contact with it as it flows through the body can be healthful, which is why acupuncture works and why doing the sensing exercise is so healthy.

If my reading of the Pyramid Text is correct, at this early time in their history, the Egyptians still had an objective understanding of this energy. Where it came from, I don't know, but it would seem that a good bit of our past is lost. (There's really no mystery about this. When the last ice age ended, the oceans inundated coastlines around the world to a depth of thirty feet and more. Underwater archaeology is costly and very difficult, with the result that we know little of what once lay along those shores.)

They saw the energy in the spine as the link between the physical body and what I see as the energetic body. They believed that there are basically three spiritual bodies, the *ka,* or nonphysical double of the person, the *ba* which was able to travel between the world of the living and that of the dead, and the *akh*, which is the part that survived death. They believed that the evil don't grow the *akh,* and just disappear after death, just as Anne observed happening after she died.

The Pyramid Text describes the spine as a serpent of energy that is linked to the body by seven smaller serpents that surround it, the *ta-ntr.* I think that those seven serpents became the *chakras*, or circles of energy that we know today from Indian mysticism, and that *ta-ntr* may have evolved into tantra, although there's no scholarly evidence for this.

What matters, though, is understanding that linkage. I have been out of body a number of times. On three such occasions, I was visible to other people, so I am quite sure that the belief that this is actually an internal state is wrong. One was Linda Moulton Howe who will attest to it, another was broadcaster Roy Leonard, who has passed away, and the third is a scientist I can't name.

I have tried many times to do it on my own but with little success. Twice when I have been taken out, and I feel that this is important, I have experienced either a sensation of something being unlocked along my spine or a shock going down it. Then I have been able to roll out and move through the world around me, remaining conscious and aware. When I was seen by the scientist, I was able to have a conversation with him. I was not able to control what we talked about, however. I was there as a messenger, saying to him that he needed to face the reality of his soul and to do whatever he needed to help him strengthen it by coming to terms with himself. He chose to go in the direction of religion. This is fine. As I've said, religions offer useful paths.

When I go out, my consciousness, or the part of it I identify as me, is what emerges. My name does not come but rather something deeper. I think that this is second body. It is not formless but bordered. I feel like a sort of orb. In other words, it is a body but without a familiar shape until I become visible, whereupon it turns into a form recognizable as me, generally in clothes I was wearing either at the time of separation or shortly before.

I would like to mention again at this point that I cannot control this. The two clearest separations have begun with that sensation of being unlocked along the spine. When I have been seen, I have in no way chosen this, nor have I been able to control the process directly, although I suspect that it has something to do with the ease of self-sensing that I have built up over years of my meditation practice.

The Pyramid Text says that that the spinal cord contains light. We can detect a concentration of electrical energy in this dense nerve bundle. Is that electricity somehow different from the light that they perceived? I suspect that the answer is both yes and no, in that the life force generates the electrical energy in the spine but is not that energy. We can detect the electricity. They could see the life force. They also believed that it can leave the body and remain coherent, and that it does so at death. So the energy involved is not like that of the physical part of the nervous system, which will wink out shortly after the heart stops pumping. This energy will instead rise out of the body as what must be a sort of plasma and begin to experience life in a new way.

I think that it is this that some of the predators want, essentially an entire life that may be tasted in exquisite detail, filled with all the energy of surprise and wonder that went into it. What I have experienced when my teachers have taken me out of my body tells me that separation can be caused by applying some sort of energy to the spine. The result is the ineffable blessing of out-of-body movement.

It might be, though, that somebody who cannot taste of the wonder of life on their own, they might want to steal it.

They might do that by ripping the spine right out of the body, thus detaching the energetic body and enabling them to capture it.

We might think of this as unspeakably evil. And it is certainly terrible for the poor person who loses this most precious of all possessions. All the effort that has gone into the life is taken, and the soul ends up empty handed. But is this evil? When a shark devours an innocent swimmer, it is terrible, but it isn't evil. It's just nature being nature. The same thing holds true when a person is attacked in the night, their spine extracted and their energetic body captured. It's just nature being nature—which is all well and good, but the immediate question that any swimmer in the ocean of life must have is, how do I continue to benefit from my swim until I am ready to return to shore without getting eaten by some soul shark?

This may seem like a very theoretical question now, and probably to many people a crazy one. But it is not theoretical, it is essential, and it is not crazy at all, but exactly as logical as how to protect oneself from a physical shark.

As they become more evident, there is going to be a lot of fear. The media will rush to tell horror stories. Believers will be trumpeting danger, claiming that these are the proverbial demons of darkest legend. Close encounter witnesses will be tearfully recounting their horror stories, many of them entirely real.

I am convinced that the main thing that has caused me to lose my fear of them is my work toward a strong soul, and I think that this is how we will defeat the fear. How ironic that, to save ourselves from the predatory side of the visitors, all we need do is to become good human beings.

Living a life of love, compassion and humility is all that it really takes. It isn't necessary to engage in religious rituals, hexes or anything like that. It isn't even necessary to believe that the visitors are real or that the soul exists. All that *is* necessary is to understand how to live this way and to do it with all the determination that one can bring to the effort.

Knowledge of why living a good life is important goes all the way back to our very oldest moral codes, to the Egyptian law of *ma'at* and the Ten Commandments that are a distillation of its many admonitions.

In my experience with the visitors, the first formal lesson I was given was one in humility. It was 1988, *Communion* had just been published and I was the king of the hill. My brother, eleven years my junior, came up to the cabin to see where it had all happened.

As I proudly took him down to the clearing from which I had originally been taken up into the UFO, I heard a low, tough voice say in my head, "Arrogance. I can do anything I want to you." I hoped that it was just my imagination admonishing me and decided to tone down the bragging a bit. But when we got to the clearing, a huge UFO appeared. It was early evening, and there was no question what that big oval disk in the sky was. We both stood there looking right up at it. Then I saw three figures in a nearby clearing. He didn't, but that didn't matter to me. I knew now that the voice had been real, and to my astonishment, I realized that I was being called out for my lack of humility. But what would happen? "I can do anything I want to you."

The next morning, my bank called. I was told that I had no account with them, and there were checks needing to be paid. But what could they be talking about? Of course I had an account! All the money I had in the world was in it, except for about fifty dollars in my wallet. The banker suggested that maybe I'd moved to another bank. I told him that they'd lost my account. He was disbelieving but agreed to search in their records for it. He also said that if the problem wasn't resolved by the close of business, they'd have to bounce my checks. I had a hell of a day waiting. He called at about ten minutes of five and told me that they still hadn't found my account. After a bit of wrangling, he agreed to hold the checks over for another day and get a search done in the bank's backup records.

During the sleepless night that followed, I reflected deeply on humility, thought long and hard about how I lived in my ego, and considered that "Whitley," after all, was just a name, and that back somewhere behind the edifice of the famous writer was a soul trying to accomplish a life task that was only impeded by his inflated self-importance. I resolved to reimagine "Whitley" as a social tool rather than as all that I am.

The next morning, the banker called. My account had been found in the backup records they kept in case of a general electronic emergency.

Humility is a task I still work on every day of my life.

Another excellent lesson I received, that is in my mind connected with them, or related to their presence in my life, once again involved the Master of the Key.

I had never understood exactly what sin was. When I was a boy, I lived in a maze of Catholic sins, all designed to make certain that the church, with its ability to manage forgiveness, remained central to our lives. We used to get little cards that showed how many years in purgatory each sin would lead to. Sass your mother, expect to burn for a thousand years. Do it to a nun, expect a hundred thousand. Eat meat on Friday, burn in hell forever.

I couldn't figure it out. And why was fish ok? It was meat, too, wasn't it? And what about chicken? Did it mean hell or just purgatory?

I could understand things like murder and robbery and such as sins, but most of us never do anything that bad. In fact, as I know now, great sin is rare. It takes work to do real evil. But doing things that we will later regret in our lives is commonplace, and it is our fear of our regrets that causes our fear of the visitors.

During the meeting with the Master of The Key, as I sat listening to one wise statement after another, I decided to ask him about the mystery of sin. He replied at once: "Sin is the denial of the right to thrive."

Since the moment I heard those words, I have been examining my life through them as best I can and also understand enough about the way others are living to be able to be compassionate, which proves to be an extraordinary challenge.

I don't think of myself as being qualified to tell anybody else how to live a moral life, let alone what compassion means and how to enact it. I can, however, say what it means to me. It is always thought of as being endlessly forgiving and kind, but it is not that. It begins with looking deeply into people, oneself included, without judgement or preconceptions, and finding what needs you can fulfill. This includes everybody, not just the people around us but also every creature, be it physical or otherwise, from the grass under our feet to the angel soaring.

One of the loveliest and, I feel, most useful things Anne said after she died was, "We are, each of us, all we have." If we really take that to heart and make it part of mind blood and bone, if we live it and breathe it every moment, there is really nothing else that needs to be known about compassion. If we put ourselves behind the eyes of any creature, no matter how humble or how great, we will see immediately that we all share the same struggle and are, each of us, deeply alone and in need.

When I was a boy, I asked one of the nuns at my school why she was a sister. She said, "Because here I am always needed."

That is true of every one of us. Know it, and compassion becomes your path.

We must understand ourselves if we are to understand others. This cannot be done with ego. It takes humility. Once we do it, then love comes forth, and we are the stronger for it. We will see what others need through the medium of seeing what we truly need. And, into the bargain, we cease to have anything about ourselves that we would prefer to hide. So when the visitors look at us, their vision penetrating to our truth, our truth can look right back, unafraid.

So love flows out of compassion, which rests in humility. As Anne communicated with me from beyond the grave with eloquence and ease, I realized that she, who was in her essence a teacher and a wonderful one in this life, had achieved mastery in the next level. I asked her to help me find an aim that would give direction to the rest of my life. This is when she said, "Enlightenment is what happens when there is nothing left of us but love."

Live that, and the visitors will cease to be demons in your view and become angels. As is said in the film *Jacob's Ladder*, "The only thing that burns in hell is the part of you that won't let go of your life: your memories, your attachments. They burn them all away, but they're not punishing you, they're freeing your soul. If you're frightened of dying and you're holding on, you'll see devils tearing your life away. If you've made your peace, then the devils are really angels freeing you from the earth."

Life with the visitors begins when we have made our peace with ourselves.

# 10

# DARK TRUTHS AND LIGHT

lose encounter is not only among the most mysterious and complex of human experiences, and certainly among the most fulfilling, it can also be dangerous. But for most people, it is not that. It is far from that.

Still, if there's any danger, we need to understand very clearly what it involves and how we can cope with it.

Some of the early researchers saw only the dark side of it, but at the time the modern experience was just beginning and there was little else to see. The forced abductions started in the 1960, so when people like Budd Hopkins became aware of them, there were no witnesses who had as yet developed relationships with the visitors, and pretty much everybody was terrified.

When I first met Budd, he was careful to avoid sharing his thoughts and opinions about the story I told him. For my part, I had nothing to say about it other than to describe it as it had emerged into memory. At that point, only Timothy and Anne had heard it and I was keeping my injuries very much to myself.

Budd suggested hypnosis, and referred me to Dr. Donald

Klein, who performed the two hypnosis sessions which led to the writing of *Communion* and are archived on my website. From the horror of those experiences to the life I live now has been a long road, but I have not traveled it alone, for it is now seen to be the path of many, if not most, close encounter witnesses.

The experiences often start violently. Great terror can be involved—and why not, who could fail to be terrified by the apparitions that we see? Some of us, most particularly those who have only a few experiences, never get beyond that point. Others report being continually hounded by bizarre entities. But for most of us, there is an arc that starts in fear and ends in a life of deep inner search, and psychological, intellectual and spiritual exploration.

During my first hypnosis with Dr. Klein, I remembered seeing one of the *kobolds* standing across our bedroom. This result was entirely unexpected. In an instant, my vision of the world we live in was upended. The thing didn't look all that menacing, but it was *there*. I erupted into a torrent of screaming so intense that it nearly brought the police. I'd never known that such an intense feeling of fear, fiery, raw and desperate. And yet it had been hiding inside me all along, certainly at least from the previous summer. My interior underworld had risen to the surface.

Reading the literature of close encounter, including the letters that Anne saved and the dozens of books by researchers like Hopkins, David Jacobs and Dr. John Mack, one sees a picture emerging that is quite like my own experi-ence: a frightening, world-shaking initial encounter, followed, if the experiences continue, by a long struggle to come to terms with the situation.

The FREE organization, founded by Rey Hernandez and sanctioned by Dr. Edgar Mitchell, is embarked on a study of close encounter witnesses designed to explore how their relationship with the visitors evolves over time, where it leads, and how the witnesses' perceptions change because of what they are experiencing. (FREE is an acronym for Foundation for Research into Extraterrestrial and Extraordinary Experiences.) Rey, an attorney, was inspired to start the organization by his own close encounter with a UFO, which unfolded in the presence of his wife and daughter. Like me, like Budd Hopkins and so many others, he had been called to action by the phenomenon itself.

Despite the usual paucity of funding, FREE has managed to accomplish a substantial amount of statistical research, the results of which are compiled into a book called *Beyond UFOs: the Science of Consciousness and Contact with Nonhuman Intelligence."*

By creating a website designed to attract the interest of close encounter witnesses and providing extensive professionally designed questionnaires, they have developed the beginnings of a profile of what people actually think about what's happening to them. Of course, what they have done so far suffers from being a self-selected sample, but nobody in this field presently has the financial resources to do random sampling. Even so, with over 5,000 respondents at this point and a high level of consistency, there is reason to believe that FREE has developed a reasonably accurate picture of the human experience.

What is found essentially agrees with the work of Dr. Jeffrey Kripal, who sees the first contact as initiatory in nature. In our book *Super Natural*, he says, "The calling of the shaman is often signaled by what Mircea Eliade, in his classic study *Shamanism: Archaic Techniques of Ecstasy* (1964), called an "initiatory illness," a severe psychological trial or physical illness that effects a transformation of the future shaman's being, that spiritually mutates him, if you will. Other common tropes include the presence of "power animals" or totems, the ability to leave one's body and travel in the interworld, a proclivity for trance states and robust visionary experience, erotic contact or marriage between the shaman and a particular deity, spirit or discarnate being, and the use of psychoactive sacred plants to catalyze and super-charge these various magical powers."

This could be a description of the opening phase of a life in close encounter. As such, I see it as one part of a vast process of re-enchantment of the world. In *The Afterlife Revolution*, Anne and I make the point that the near death experience is, like the close encounter experience, upending of one's understanding of reality and thus also initiatory in nature. In addition, medical advances which are increasingly able to return people from the edge of death, are causing a distinct increase in such experiences. In *Changed in a Flash,* for example, Elizabeth Krohn describes actually being asked to decide, while her body lay effectively dead, whether or not to come back. She did, and was able to because modern medical science could save her.

What we would appear to be looking at, then, in both the close encounter and near death experiences, is a massive increase in initiation. So this is not so much about the arrival of aliens as it is about a change in awareness—essentially, a deepening of the human experience of reality. These experiences challenge the way the mind sees the world, and even change, along with meditative practices, how the brain works.

While there is an expansion of consciousness in process on a massive scale, I personally cannot ignore the findings of Hopkins and Jacobs. This is because the horrors they describe in their books happened to me and Anne. During my abduction, a device was inserted into my rectum that caused an erection. This wasn't anything exotic. It was an electrostimulator, used in those days in cases of sexual dysfunction and still common in animal husbandry. My frantic effort to push it out was what caused the rectal tear that left me in pain for years thereafter.

In my second hypnosis session, you can hear me reliving the experience and commenting with confusion in my voice that I have an erection. Semen was then extracted from me. Sometime in the next year, Anne and I had the experience of being shown a baby, exactly as Hopkins and Jacobs report.

So, as always, this experience has more than one level of complexity. On the one hand, it is certainly leading toward some sort of awakening of the human species. On the other, there are indeed ominous, dangerous and bizarre elements that should not be ignored.

Neither, however, should they be dwelt upon to the exclusion of the equally important fact that there are benefits being gained from close encounter.

I know that what I am about to describe will be seized on by sinister forces who want to assert power over others through the use of fear. Because of this, and because fear sells, the media will also rush to it, as will all the conspiracy theorists and paranoids who make so much noise in relation to this matter. But if I don't include it and the story comes out another way, then my whole effort to build a basis for communication between us and the visitors could go to waste.

The story I am about to tell is a true one. Of that I'm reasonably sure. But is it about the visitors? Of that I am not at all sure.

I am not an advocate either for the idea that they are benevolent or exploitative, good or evil. My experience of them suggests that they are more complicated even than we are, and the moral range in human society is very broad. At any given time, our species contains madmen, criminals, saints, and, for the most part, a broad cross section of average people with all their ordinariness and imperfections—and their excellence.

I cannot imagine that the visitors, who may not even be entirely of this world or all stem from a single evolutionary background, could present any less ambiguous a face toward us than we do toward ourselves, and, for that matter, toward them. In fact, if we went to another world, its inhabitants would see us as a complex, contradictory presence. Religious groups might go, scientists, tourists, all sorts of people, including the insane and the criminal. If you add to that something that we will soon see among us, which is intelligent, even conscious, machines, then the diversity and contradictory nature of what the innocent locals would experience would probably be very similar to what we experience now.

Missionaries would have one agenda, anthropologists another. Biologists might abduct the locals in the same way that we do wild animals and the visitors do us, in order to extract their DNA and sexual materials so that we could understand them better. As we also do in the animal world, they might breed examples of us, in part to preserve us and in part to study us. If there was something about us of value that could be harvested, that might be done, too. If it was illegal to do this, it might be done anyway, although by criminals and thus on a smaller scale than, say, officially sanctioned abductions.

In other words, our relationship with innocents in another world might look very much like the visitors' relationship with us.

Just as a session with the vet is terrifying to a housecat, a session with the visitors is terrifying to us. We are not house cats, however, and we can learn more and come to understand their motives, methods and aims. I believe that we can also cause them to recognize the dignity of our being, and perhaps this has been happening over the past forty or so years, which would explain the steady decline in horrifying abductions and the increasing sense of relationship with the visitors that is growing among close encounter witnesses.

By communicating coherently and forming relationships, I feel from personal experience and from the experiences being reported by so many others, that we can individually and, I would think, probably also collectively, improve our standing with them.

Animals may be killed by human beings without much, if any, constraint. When they are sick or abandoned, we euthanize them or simply let them die. They have rights in only a few human societies, and those are limited. We carry out research on them that causes them suffering and can kill them. We can take them from their families and not return them. We can keep them in cages for our amusement until they die.

So when we see what the visitors do to animals, such as the infamous cattle mutilations, we are not seeing anything that we don't do to them ourselves. Their methods are just different, which is why we find them disturbing.

Since at least the beginning of the 20th century, farm animals have been the victims of a bizarre form of mutilation that involves things like the removal of eyes, the cutting out of tongues, the shearing away of lips, the coring of rectums and the draining of blood. They are also often found with their spines drawn out of the vertebra, which would seem to be impossible without splitting the bones. Most recently, a number of these events were reported in Oregon in July of 2019 when 5 prize bulls were found showing signs of this mysterious sort of attack.

The media, when they report these events, generally take the word of local sheriffs that it is predator related, although National Public Radio reported the July 2019 mutilations as more of a mystery.

There is a distinct possibility that what happens in the dark of night on isolated ranches is not normal predator action. Among other things, the spinal cord cannot be removed by a coyote or a mountain lion, not without smashing the spine itself, vertebra by vertebra. The carcasses are usually found with all the blood gone. A coyote isn't going to drink every drop of blood, leaving the carcass and the ground both dry.

To anybody who is aware of the significance of the spinal cord as the connection between the physical and energetic bodies, this would be an especially worrying aspect of the phenomenon. We don't like to think of animals as being conscious or having souls, but, as my cats revealed so eloquently with their fear during the night of the nine knocks, soul is everywhere and it can be made to feel vulnerable, at least among the higher animals.

Could it be, therefore, that the spines are pulled out in order to enable the energetic body to come free and be captured? In my life, the unlocking along the spine isn't threatening. On the contrary, when it happens, my reaction is excitement.

I am not so sure that it is always like that.

Not only sheep and cattle, but pet cats are also the victims of mutilation. It makes me wonder if my cats didn't somehow intuit that, which would explain their terror that night. In July of 2015, Linda Moulton Howe, who is the world's leading expert on this bizarre phenomenon, reported on Dreamland that the cats are found with "very precise cuts. Some cats are just cut in half, like with a band saw or very sharp knife or something like that, but no blood. Just cut in half and either the front half or the back half is left to be found. Other animals have only flesh removed from the abdomen area, or a few organs removed, or all organs removed. A few have had just the spine removed in a very precise way. Usually there is no blood." Some of the cat mutilations have come in waves, moving from city to city around the world, as if somebody very unpleasant was slowly circling the planet committing these murders in an organized and methodical way. As recently as August of 2019, cat mutilations were reported in Everett, Washington. On August 10, 2019, a representative of the Everett Police Department said on KING TV, "These are very unique injuries that do not appear to be caused by another animal." As always, no perpetrator has been found even though, in this case, all five mutilations took place in a single neighborhood. All that remained were legs, uniformly bloodless.

This is only the latest of many such cases worldwide, most of them involving cats cut in half and their bloodless remains left where they had originally been picked up. Like the mutilations of farm animals, despite extensive investigations in city after city, no perpetrators have ever been found.

Why aren't they, though? Could it be because the perpetrators can read and control minds and, therefore, are impossible for us to catch? If I had not observed such abilities in action, I would never believe that such a thing could be possible. But I have seen this. When Anne and I were living in our small condo in San Antonio and the bizarre man who had been living in the woods behind our cabin in New York showed up with two companions, bizarre events immediately followed, most particularly clear examples of their ability to control minds. He was short and appeared to be a feral child. He smoked constantly, which is why I originally noticed him in our woods. I was concerned about all the smoking in a dry August and approached him to caution him. As I did so, it became obvious that this was no ordinary child, if a child at all. I left the area.

After we lost the cabin, we moved to a small, ground-floor condo in San Antonio which had a screened in porch that opened onto a garden and a cul-de-sac beside it that created a shadowy space just outside the bedroom. No sooner had we settled in than I realized that he was standing in that cul-de-sac at night, smoking constantly. I found that I could feel him inside my mind, literally sense another presence in me. This was nothing like communing with the visitors. There was nothing gentle or supportive about it. In fact, it seemed somehow sexual, and in an ugly, invasive way. It was nonphysical rape, to be frank, and made me extremely uncomfortable. I felt explored in some very private parts of my mind. Looking back on it, I can still feel the curiously thrilling and yet ugly sense of it. Looking back, it reminds me of the sense of domination I felt after the communion experience, that led me to write the short-story "Pain."

This is what, in so much human mystical tradition, contact with the dark side entails. It takes you into places in yourself that you otherwise would never go, but, once you are there, you find your own darkness, and its mystery and the thrill of it. In my case, when he entered me the way he did, I experienced homoerotic pleasure. I was afraid of him—and of the part of me he could connect with—and I drove him off. But that is in me, too, just as is the erotic masochism that I explored in "Pain." When Anne read that story, she said, "This sounds like you want to be whipped. Great, let's get started right now." She would have done it, too, but I did not dare to take the experiment farther. Perhaps I should have, and perhaps I should have let that dreadful being enter me more completely, but I did not dare to do it. When we invite our own darkness into our outer lives, there is no guarantee that it will stop when we want it to.

As the nights passed and he lingered just outside our bedroom, just a few feet from me, Anne and I became aware of two very strange men living in the flat immediately behind ours. He was living with them. One day, I was in the local drugstore when I saw one of the men loading shopping bags with smoking materials of every kind, which were in those days still sold on open shelves. Anyone in the store could have seen him doing it. He walked out in full view of the clerks with two bags full of cigarettes, pipe tobacco, cigars, you name it. Every clerk in the store stood silent, staring straight ahead. Except for me, the customers were all in the same condition. As he passed me, he gave me a look that was at once knowing and venomous, and from that moment to this, I have known that there are people who can do as they please in this world, because if you can control the minds of the people around you, you can control your world.

I soon discovered that the three of them were squatting in the condo complex. I told this to the owner of the condo they were crashing in, who had them evicted. The last I saw of them, the two men were canvassing the complex trying to sell the rest of us the owner's furniture. Of course, as everybody knew they were squatters, nobody bought it. One morning a couple of days later, the "boy" strode out of the space between our condo and theirs while I was working in our garden. He went marching off down the street.

And this is why I know there are people who can control the minds of others. One would think that this would confer on them almost unlimited power, but one glance at the seething, desperate world of the human elite, and it is obvious that, whoever they are, they do not rise to the top of our societies. Judging from the way these men were, so very weird and, in the case of the small one, apparently schizophrenic, one can see why. Nevertheless, this part of my experience has been richly productive in terms of personal insight. Because of the erotic domination involved, I discovered aspects of myself that needed to be brought into the light and accepted as part of me. In this sense, what happened to me can be looked upon as therapeutic. And this is the way of the dark side: it is in darkness that we discover what needs to be brought out into the light.

I can see where beings such as those three might visit their attentions on housecats. If the ones we encountered are at all exemplar, they are seething with vindictive menace and so might destroy the cats simply because they are loved.

In the year 2000, I came across some cases where human beings appear to have undergone the same fearful mutilations that have been visited on animals. Until recently, I didn't have much to support this other than one equivocal case from Pennsylvania and a second-hand report from New York of a number of exceptionally brutal unsolved murders, all of homeless people. Recently, though, I obtained a transcript that suggests that the cases should be taken seriously.

What I initially heard were two stories. The first, in 2000, was that a total of seventeen street people, all without known relatives or anybody who would really care about them, had been taken from Brooklyn, and possibly other cities in New Jersey and New York. I was told that they had been mutilated by having their eyes, tongues and genitals cut out while they were still alive, drowned in the ocean then left on roofs near the places where they had originally been kidnapped. At some point, a three-centimeter incision had been made just below vertebra C1 in the spinal column and the spine somehow pulled out through it.

There was also a case in Pennsylvania in August of 2002 that I was briefly involved in at its outset. I first heard from Peter Davenport of the National UFO Reporting Center that an unidentified person had reported seeing an individual being lifted into the air from a woods above his farm and that the man had disappeared into what looked like a flying saucer. I told Peter that, if this was true, there would soon be a missing persons report—as, indeed, there was. A search was mounted for a man called Todd Sees, who had last been seen riding a four-wheel drive vehicle in the woods in question.

I then suggested to Linda Moulton Howe that she go to the area and contact the local sheriff, which she did. She was warned off and told to leave town, which she also did.

About twenty-four hours after he was reported missing, according to news reports, Mr. Sees' body was found "emaciated" in a wetland a few yards from his home. There was no definite cause of death ever reported. As far as I know, no member of the family saw it. I have been unable to determine the fate of the autopsy report, so that is where this tragic case stands.

I have now obtained the transcript of an interview with one of the coroners involved in the New York cases that does show one important similarity with the cattle mutilations, which is that the spine was also observed to have been removed, although—and I think that this may be important— not with anything like the precision of the cattle mutilations.

This transcript was generated by a nurse who was talking to a coroner who was involved in the autopsies. I do not know how many of them this particular coroner participated in, and this conversation covers just one of them. Aside from the overall finding of mutilation, there were three strange things noted. The strangest was probably that some tissue from the corpse could not be identified at all. It was neither human nor animal, and yet the corpse appeared to be human both externally and internally. There were small metal balls found in the abdominal area. At the suggestion of the nurse, who was aware of research being done by Dr. Roger Leir in California at the time, where strange fluorescence was being observed on the bodies of people claiming to have been touched by the visitors, the coroners applied ultraviolet light to the remains and found that they fluoresced. The fluorescing material that was gathered off the skin was also tested in a forensics lab. The test returned as a nonorganic substance, unidentified.

At first the transcript seems to say that a small incision had been used to cut the superior transverse scapular. It develops that the conversation was actually about the transverse process, which is a small projection on each side of the vertebrae that enclose the spine.

After that discussion, the transcript continues as follows:

Coroner: We found severe damage to the muscles.

Nurse: Did it look like the spine was ripped out or cut?

Coroner. Ripped. Reason for the damage.

Nurse: OK.

Coroner: Entrance and exit damage.

Nurse: One hole?

Coroner: It looks like a wire coat hanger did the damage. A device with a hook.

Nurse: Still, isn't the transverse process longer than three-centimeters width?

Coroner: Yes, it was totally fractured.

Nurse: Were any vertebrae left? Or was it C1 to sacrum?

Coroner: The strength to pull that out…

Nurse: Were there any remnants left?

Coroner: Two small coccyx (e.g., coccygeal vertebrae). They were crushed.

Nurse: Was there exterior bruising to account for the crushing or only internal?

Coroner: Only internal. That's what's so puzzling.

Nurse: Were the abdominal organs harmed in any way?

Coroner: Not that I remember.

Nurse: That makes no sense.

Coroner: Peritoneal wall not ripped, but six ribs broken and lung damage.

Here the transcript ends.

This body was found on a roof, terribly damaged and with the spine torn out. The person had been mutilated, drowned, then suffered the additional brutalization. The victim was a homeless person with no known relatives and no identification.

Shortly after the conversation transcribed above, the corner ceased having discussions with the nurse, and they have lost touch.

This is not the only report of a spine being pulled out through a small incision near the top vertebra in the neck of a human being. So one has to ask, if the spine is removed, is the energetic body forcefully separated from the physical, whereupon it then becomes controllable by whoever released it? In short, can it be captured?

Some ancient traditions suggest that there is such a thing as a hungry soul. These are nonphysical beings who cannot enter the physical world and hunger for the taste of it—just as it appears to me that our visitors do. Such creatures would certainly have a motive to capture the energetic body, if indeed it records every detail of the life lived. It could perhaps be used like a sort of food made up of experiences, perhaps providing an imitation of life which, to such an entity, would be the equivalent of an addictive drug, almost impossible to resist. Good and bad wouldn't matter, I don't think. What would matter would be the taste of life, every bit of it.

In his book *The Cosmic Serpent*, Jeremy Narby explains that, among Amazonian shamans, there is a belief that certain spirits hunger for the taste of life, but the only thing that they can still detect is tobacco smoke. To communicate with them, a shaman will smoke powerful tobacco.

Could it be that there are physical beings who can no longer taste of life but long to do so? If they are actual, physical creatures and not spirits at all, or can manifest physically, maybe that explains the obsessive need for tobacco of the creatures who followed us to Texas. If they are living in a state of profound anesthesia to life—in effect, being dead while still in physical bodies—perhaps they have a motive for stealing a person's energetic body. Or maybe it has nothing to do with the energetic body but is done to harvest spinal motor neurons, or for some other reason involving the use of spinal tissue.

Before we let our imaginations run away with us—or rather, I let mine run away with me—It cannot be forgotten that the tales of human mutilation are only stories. A transcript of a conversation isn't a notarized coroner's report. True, I do know the nurse involved, and I don't think she's making anything up. But she is also a member of the Mutual UFO Network, which has at least a partial exposure to people within the U.S. intelligence community, which does seem to me to have been in the past interested in controlling the contact narrative with fear and, for all I know, may still be interested in doing this. Could she herself have been duped? Or given the roughness of the way the human bodies were treated in contrast to what is seen during the cattle mutilations, was this done by a serial killer?

Unfortunately, the matter has to be kept in question. There isn't enough information to do anything else. However, the FBI was involved in both the Sees case and at least the one case the nurse is aware of in Brooklyn, so something more than ordinary does seem to be amiss here.

If the removal of the spine is a way of harvesting the second body, I would then call this not only a conventional crime but also a spiritual crime of a kind that we do not have in our world, but which we had better learn about and learn to protect ourselves against, and the only way I know of to do this is, as I have discussed, by building a strong soul.

In the past, the violent side of our relationship with the visitors has often been ignored or covered up. UFO investigator Philip Imbrogno reports that Dr. Hynek would not allow reports of cattle mutilations or human abductions and deaths to be included in *Night Siege.* Dr. Hynek feared the adverse publicity that abduction reports would generate. He probably feared that the entire book would be discounted because people would be unable to face these stories. As I was personally abducted in 1985 about thirty miles from the area where the Hudson Valley sightings were taking place, I can attest, in this case, that at least one such abduction did happen.

Famed investigator of the anomalous John Keel says that the phenomenon, which he does not regard as an alien intrusion, has a long history of hostility to mankind. Dr. Jacques Vallee, who also sees it as something far stranger than alien contact, has explored this in his books, most notably *Messengers of Deception* and *Passport to Magonia.*

Although the life experience collected in the energetic body seems to me to be the primary focus of the phenomenon, the physical aspect is very certainly there, and physical contact can clearly be dangerous, at least when some entities are involved. Also, it would appear that, while much of the phenomenon is ancient, some parts of it must be quite new, at least, if the stories of mutilation and spinal extraction are true. We have no historical record of murders involving such mutilations, and if they had been going on for a long time, surely there would be a history of it. Stories of deaths involving such a very unusual injury would be famous, like the tales of Jack the Ripper. There is only one, though, prior to the early 2000s when this smattering of equivocal reports appears. Then it ends. It goes no further. It might be possible to cover it all up in the United States and some of the other developed countries, but not everywhere, and, except for one rather shaky case in Brazil, there are no other reports. This case took place on September 29, 1988. A body was found near a reservoir south of Sao Paulo that was damaged in ways that were similar to cattle mutilations and seemed similar to the cases reported in New York. But it could also have been that the man suffered chemical burns or even conventional scavenger predation after death. While photos exist, and they are graphic, they are not definitive.

Since 1985, there have been abundant occasions when the visitors could have done anything they wanted to me. Given their power, the men at the condo could obviously have done much more than they did.

I have intentionally opened myself to the visitors, without placing any limits on who might show up in my life or what they might do. Most recently, in February of 2019, when I was at a place in Texas where I have encountered them so many times, they showed up physically. I caught only a glimpse of one of them, but I could hear them very clearly all around me. I was a bit scared, but not all that much. After our quite intense meeting was over, I went on to bed, something I never could have done in the old days.

In all the letters that Anne retained from the *Communion* days, only one specifically mentions a death in connection with a close encounter experience. This involved a man who ran after some of the *kobolds* with a shotgun. He was found dead beside a small lake on the family property. His body was unmarked, but there was a bulge on his chest caused by something under his skin. When there was an autopsy, the object was taken by the authorities. No specific cause of death was found, and it was classed as a death by misadventure. The object was not returned to the family.

With our limited understanding even of what the overall

presence is, let alone how many different forms are involved and how they relate to each other, we cannot draw any conclusions beyond saying that some situations have arisen that are weird and suggestive of violence and that for the most part contact appears to be frightening and sometimes violent but usually not dangerous in any permanent way, and, if one keeps at it, it leads to the benefits of intellectual, emotional and spiritual growth.

If human mutilations are happening, that needs to be part of the public knowledge base, along with all we know about specific cases and any theories about why it happens and who is doing it.

So what is to be done? That's where communication comes in. If we are under physical threat, we need to understand why, or we will never succeed in countering it.

That's liable to be a long path, but if we are strong and committed first to our own needs, it is possible to follow it, I would think, to a satisfactory conclusion. I believe this because it has been and is my life. I am here writing this, not ripped apart in some field somewhere, and there have been and are ample opportunities for them to do that to me. I do not protect myself, not with weapons, not with prayer or ritual or anything at all. Instead, my primary focus is how to open myself more profoundly, not close the door on them and try to hide or fight or channel the relationship in any way. What they do with me and to me is entirely up to them, with the result that they lavish knowledge and experience on me, enabling me to lead a life of great richness and wonder.

So I go on meeting with them, letting them into the most intimate levels of my life and learning from them all I can.

We are alone with them on this little, dying planet. They do possess secrets that can help us, of that I am quite certain. My focus, thus, is not trying to fend them off but rather to cautiously take what I can of what is on offer from them, which seems by far the more practical and useful approach.

It's not hard to believe that cases in the developed world could be covered up by some sort of concerted international effort, but if murders this bizarre are happening worldwide, some cases are bound to have come to light, and that has not happened.

Given this, I think, on balance, that we're going to have to table our caution until there is more justification for it and do what we can to get our relationship with them to come into focus. I think if people begin announcing their availability, say with the sensing exercise, there is liable to be a response. This might be doubly true with the intellectually accomplished, as I know that they want particularly badly to communicate with scientists and others who can help us save our planet.

If they are going to emerge, we're going to have to accept certain basic realities. They are that we don't know what they are, that we don't know the degree of danger involved and that we are going to have to take a gamble. But it is an informed bet. If they were an invading force, we'd be their slaves by now, or dead. While not all of them may have motives that are to our liking or in our interest, some of them do, or I and others like me would not be having positive experiences with them. When I reacted to them by trying to overcome my own fear, they responded in a deeply positive and persistent way. They demonstrated with Dr. Belbruno what is on offer to our science and with me what is available in terms of a richer and more true inner life and understanding of the world.

Still, it's going to be a very hard job for us to do this.

How can entities with different biology, different histories, different perceptual systems and a different relationship to reality possibly hope to make any sense, say, to government officials or scientists who cannot conceive of any form of communication other than the spoken or written word, which are in fact, not adequate to the situation? There is simply no way we can tell what is actually happening when we attempt ordinary conversation with them. They may choose words, for example, that they have learned will be a fitting response to some question or comment, but how can we know what they think they mean?

When I was with the visitors in December of 1985, they used a machine to generate a gentle, feminine-sounding voice that kept repeating, "What can we do to help you stop screaming?" They obviously knew that the words and tone were intended to communicate reassurance. But did they know their meaning? The voice was gentle but also lifeless. There was a definite sense that it was machine generated and, thus, very far from reassuring. If anything, it only increased my fear. I must also note, though, that they did not want me to scream. They wanted me to be calm, and if I had been, I think that things would have been much less fraught. But given what I was seeing and what was being done to me, that was just not possible.

Ordinary language is unlikely to be a reliable tool, as both sides are unlikely to know whether or not they are communicating useful meaning when they are using it. I can see that the use of demonstrations, such as what we witnessed at the cabin when the man from paradise expressed himself, could be a valuable basis for communicating with our academic community, where the symbols and archetypes that underlie the intellect have for years been the subject of study. The work of Joseph Campbell and Carl Jung are examples of this sort of study, and I can hardly begin to imagine what depths of insight those men might have reached if they'd had access to the man from paradise.

So we must leave the physical danger case open until

there is more definite information. Even if human mutilations are taking place, they are not a major factor, or they would be better known. Even if physical danger is not all that significant, and there is no more esoteric danger to our souls, the visitors themselves have drawn attention to another danger, and it is a true one, that must be addressed with the utmost care by every level of our society.

# 11

# THE DONATION SITE AND A WORD OF CAUTION

I f this book serves its purpose, what has already happened to a few scientists and academics is going to become more widespread. Ed Belbruno is not the only scientist they have contacted. Another friend, the Nobel Prize winning chemist Dr. Kary Mullis had a marvelous encounter with one of them at his vacation home in northern California, which he wrote about in his book *Dancing Naked in the Mind Field.* There are others, but these are two very prominent ones who I know personally and who have gone public with their stories.

Expanding contact is essential, obviously, and now that this is possible, I want to draw attention to a warning about it that the visitors left behind in a field in New Mexico many years ago.

In *Jesus Thaumaturge*, Bertrand Mehust extensively discusses the bilocation of a French nun, Sister Maria Fernandez Coronel. Also known as Mary of Jesus of Ágreda, her story is particularly relevant here. Starting in 1620, she allegedly began to be seen by the Jumano Indians in what is now New Mexico. She instructed them in Christianity and even brought them rosaries. They contacted Franciscan missionaries and told them of the lady in blue who had taught them the new faith. The miracle has been studied by Catholic investigators but never confirmed because there is no unequivocal evidence. However, whether or not the bilocation actually took place isn't important here. What is germane is the cultural dislocation that the Jumanos endured and the place where it unfolded.

The donation site where a UFO crash occurred years back is in the same immediate area that the Sister Mary story unfolded. The site is so named because the scientists who have harvested debris and artifacts from it for years for study regard what crashed there as a donation not an accidental crash.

The fact that the donation site is in an area where a seemingly impossible event took place in the past that was an attempt to minimize the effects of a dangerous cultural upheaval is important.

The Jumanos believed that they were given a new spiritual path by mysterious means, and the donation site is offering us something similar—a fundamental increase in knowledge that includes not only science and technology but greater insight into what it is to be human—and for that matter, to be alive at all.

If the materials from the donation site could be released into the general scientific community, it would change the world. And academics, by analyzing the relationship between what happened to the Jumanos when the Spaniards appeared and comparing it to what is happening as the reality of the visitors comes into focus, will safeguard us from falling into the same trap that destroyed the culture of the Jumanos.

To understand the warning, we need to look back to what was happening in the region in the early 1600s. The Spaniards were in the process of expanding north out of Mexico, bringing with them a new social order and along with it disease, disruption and the Inquisition. They had already invaded Mexico and were completely destroying the native civilization and enslaving the Indian population. By the time a hundred years would pass, the great majority of the native people would be dead, and the rest would be slaves, trapped by the Spaniards in a horrific servitude from which their only escape was death.

The placement of the donation site in that particular location is to me a clear message: contact will bring rewards, but it is also dangerous. As it unfolds, we have to be very careful to stand by our own beliefs and expectations for life and to treasure ourselves and our civilization. We need to present our visitors with an open mind but also a careful one. Otherwise, the warning is clear: things are not going to go well. Spiritual and mental contact are real and are deeply shocking. But physical contact, which is what the donation site is all about, is going to be even more intense. It will bring with it rapid cultural change, confusion, fear and all sorts of unanticipated consequences. But it is on offer, or the donation site would not exist at all.

The fact that it is where it is suggests to me that we are being warned that we might encounter entities whose culture will overwhelm ours, so we must keep our own counsel, make sure that our society remains intact, and stand up for the value of our minds and the cultures we have created. Unless we take responsibility for ourselves and our part in contact, we cannot succeed. To me, successful contact means three things: first, clear communication; second, enrichment for both sides; third, life shared in the new way of communion.

But it also offers another view of one of the continuing themes in this book, which is that care must be taken. Not only must we be careful not to become suppliants when faced with breathtakingly advanced skills and technology that we long to possess, we must also be aware that the entities that possess them, and the extraordinary power that accompanies them, are not necessarily going to also be more ethically advanced than us. The Spaniards were technological masters when compared to the native peoples, but they were also moral primitives. They had no respect for the cultures they were encountering. Everywhere they went, they destroyed the local culture, enslaved the population and killed with wanton abandon. Indigenous cultures all over the world have been shocked in this way, in many cases shocked to death, by the appearance of western civilization, often even when the approach was relatively benevolent. The reason is that, when people placing their trust in certain beliefs are confronted with what to them appears to be a far greater power, they lose faith in all of it: their gods, their beliefs, their sense of self-worth. The cargo cults of Melanesia are an example of an attempt by a less powerful culture to acquire the desirable qualities of a more powerful one but without understanding anything about the skills that are actually involved. There is a level of humiliation and a loss of self-worth that leads to the degeneration not only of the secular society but also the spiritual culture, even when it is actually superior to that of the more powerful technological culture.

An example of the danger that we face: aliens appear here who can traverse the vast distances of space. We are completely overawed by their accomplishments. When we begin to communicate with them, they say that there is no such thing as the soul. They say that not only are our religions fantasies, but that the very idea of spiritual life is as meaningless as so many of our intellectuals believe.

Or if they say the opposite: they announce that there is a god and they are the angels.

We are going to be just as vulnerable to either pronouncement, or any pronouncement they may make. The moment we become aware of their powers, we will be in danger of feeling ourselves, our understanding of reality and our abilities inferior to theirs.

In short, we will feel disempowered, and just as that feeling has been an illusion for indigenous peoples all over the world, it will be one for mankind as a species. When we look for the rewards of relationship with them, we must also look to and embrace our own strengths rather than becoming obsessed by their skills and their possessions.

It will help us also to remember that, while their skills may be far greater than ours, their vulnerabilities, while different, may be profound.

Let me explain.

Any ethicist will tell you that technological and ethical advancement do not necessarily go hand in hand. The briefest glance at our own history reveals that the Spaniards were hardly the only society to trample on less technologically advanced cultures. An even more egregious example is Nazi Germany, which used its superior organization and advanced weapons technology to lay waste to most of Europe and senselessly annihilate millions of innocent people. And why? Because the entire culture of Nazism wasn't just ethically impoverished, it was proactively insane.

So the donation site stands as both a promise and a warning that it's a big universe out there. We have obtained marvelous technology from it. Jeff Kripal has gone so far as to identify one of the people who has done the most in terms of transforming what has been found into beneficial advances for mankind as an angel.

It is time for a new conversation with the visitors to begin. The challenge goes both ways. They have to take the risk that we won't be able to bear their presence and we have to take the risk that they might be dangerous to us in irrecoverable ways.

There is such promise for us: the knowledge that the energetic body is real and that it exists to serve a soul that is also very real, the knowledge that you can leave your physical body and travel far, the promise of learning how to accomplish reliable communication between physical and nonphysical mankind. Most of all, gaining real communication with the visitors and from there entering a state of communion with them that enriches them with newness and us with knowledge.

As we do these things, I think we will see that a lot of what looks like extraordinary technology—the ability to be invisible, to travel great distances instantaneously and without a machine, even the ability to cross between universes—is really something intrinsic to life itself and which can be learned. I think that such abilities are physical expressions of soul craft.

Among the most important advances that are likely to take place as we get closer to the visitors is the institutionalization of nonphysical travel.

Because second body cannot be detected by the instruments we now possess, the idea that second body exists, let alone that such travel is possible, is generally rejected.

When I was last seen out of body was at the Institute of Noetic Sciences in northern California, by the scientist I cannot name. I was in a dormitory of small single rooms. I'd meditated as usual at 11 and 3 when, to my annoyance, I was woken up at 4 by what felt like an electric current shooting down my spine. I got up, tried to meditate again, but was too tired then went back to bed. The next thing I knew, I was in the hall outside my room. I turned around and saw that my door was closed and therefore also locked. For a moment, I thought I'd locked myself out but then realized that, because of the gliding motion of my body, I was, in fact, in the nonphysical state. I thought immediately that I would like to make myself seen to as many of the academics and scientists at the conference as I could. I went into Jeff Kripal's room, but he was sleeping like a log. I didn't try to wake him up. I know from experience that I can't do anything physically in this state. I've tried, for example, to pull a leaf off a tree. Although I myself am woken up by nonphysical entities all the time, so far I have not been able to affect anything in the physical world while in a nonphysical state.

After trying Jeff, I tried the person in the next room. The result was the same.

In the third room I entered, I found the occupant sort of half awake. I did the sensing exercise, which, I believe, may help make me visible. It doesn't feel like it does in the physical. In this state, it is like becoming aware of one's boundaries, as if you are somehow cupped in your own hands.

He saw me all right. I could see it in his face.

I found myself conversing with him, but not of my own free will and not physically. I was being used to deliver a message to him about the soul and the importance of living a life that leaves one unburdened by guilt. Some months later, he chose to take his soul's journey on the path of religion. If one treads it without becoming doctrinaire or falling into the trap of belief in such a way that one feels compelled to hurt others, it can be a rich experience.

After the conversation ended, I found myself rising. I went up through the ceiling, ascending very quickly, so high that I could see the line of dawn to the east and the dark shadow of the California coast below. I was at the edge of the atmosphere, and it was glorious. I was neither hot nor cold. I didn't need to breathe. Then I shot down and eastward at breakneck speed until I found myself on what appeared to be a college campus. I noted all the details I could. I tried to make myself seen by a man on a sidewalk but failed. I went into a dormitory and down a hallway but could not rouse the man I found sleeping in one of the rooms. I saw something leaning against the wall beside the door that looked to me like an exceptionally large dulcimer. Later, when I entered that same dorm in the physical, I saw that it was a type of skateboard that I hadn't known existed, a longboard. This tiny detail should not be forgotten, because it represents a much larger issue about what is available to one when not in the physical. I find that I have a hard time correctly perceiving something that I had never seen while I was in my body. Without the comparative files and logic of the brain, it seems very hard to do that, at least for me.

A few minutes later, realizing how far I was from my physical body, I felt a desire to return to it. I floated gently upward then shot back to my body in a flash.

I lay in the bed in a state of amazement. What an experience!

The next morning at breakfast, to my delight and amusement, I found that the scientist was telling people what had happened, and that he had seen me go up through the ceiling of his room. When I described the campus I'd been to, Diana Pasulka and one of the other participants knew it well: it was theirs, the University of North Carolina at Wilmington. A year later, I found myself in the amazing position of going physically to a place that I had first seen while in the nonphysical state. When I entered the dormitory I'd gone into in the nonphysical state, I found myself looking at what may have been the same longboard I had seen the year before but this time understanding perfectly well that it was some sort of skateboard.

While some members of the scientific and academic communities may be more willing than in the past to entertain the possibility that unidentified flying objects might be a real phenomenon, it remains strictly forbidden to discuss the idea that something might be coming out of them and that people might be interacting with them.

This position is obviously illogical, but it nevertheless is widely held. Serious scientists, serious academics and the serious media will not discuss the likelihood that something is liable to be in those UFOs—or more accurately, I think, will not face it.

Given the urgency of the situation, in February of 2017, I begged them to emerge, persisting day and night to the point that they got angry at me. I was convinced that my own example proved that our fear reaction would not overwhelm us. As I had already balked at direct, physical contact twice in my life—the nine knocks and the incident in the woods in February of 1987—they conducted an experiment.

I woke up in the middle of the night aware that there was something in the bed between my legs. It was lightly touching my genitals. As I live alone and have no pets, I immediately concluded that it was one of them. A terrific shock of terror went through me, and I leaped out of bed. So did the dark shape, which shot up toward the ceiling and was gone. I was left with a deep scratch on my left calf that was still faintly visible a year later. I sent a picture of it to the group of people I work with, with a description of exactly what had happened.

Then they embarked on what has amounted to another wonderful lesson, attempting to reveal to me what they want, what they have to give in return and how communication can work. This book is the result.

As I have said before in these pages, the visitors are a complex presence...just like us. Because we know them so little, we tend to think of them as alien races. Each species is either good or bad. Belief in evil reptilians and angelic blonds are two examples of this. But how can that possibly be the case? Look at us. Adolf Hitler and Mahatma Gandhi lived at the same time, and that is just one example of the tremendously varied array of lives, motives, beliefs and cultures that characterize just one species—our own.

Only if we can open ourselves to the possibilities and dare the dangers are we going to succeed in the endeavor of contact. Because they are here to share our surprise and our discoveries, we are never going to be sure of anything about them, not even when they are more engaged with us than we can ever be with each other. This will be more intimate than that. It will be the most intimate thing that can happen. Can we get used to it, in the end, make it our own? In other words, can we bear it?

That is the fundamental question of this book, and, I feel rather sure, of human life at this time. Can we?

## 12

## IS ANY OF THIS REAL?

We have come to the core. I think that the material I will discuss now probably needs to be taken into careful account if there is to be more direct communication with the visitors. I am hoping that this book will cause them to emerge in a more definite way, and I think that this part of my discussion is what is most essential to that happening.

How ironic that it isn't proof that they are real. In order to communicate usefully with them, one would think that things like whether or not they are fundamentally a physical presence would need to be firmly established.

Except that may be exactly what we must *not* do, not if we want to develop a real relationship with them.

As we approach the coming climax in our history, these entities, which have been considered legend, myth and folklore, are apparently turning out to be much more than that. This is happening as we come closer to two things: the failure of our planet's ability to support us and proof that parallel universes are real. Does this mean, then, that the inhabitants of enigmatic places like the mirror universe are attempting to find ways to concretize themselves in our world, perhaps in order to give us aid? I think that this is one direction in which the evidence points. At the same time, fairy folk, sylphs, trolls, *djinn* and even to some extent what in the past were called gods are turning out to be real entities that have probably been with us right here in our own universe from time immemorial, which we now perceive as aliens. Add to this the addition of a previously unremarked presence that started mutilating cattle on an ever-growing scale about 150 years ago, and may recently have done the same to a few of us, and you have quite an astonishing picture. But it doesn't end there. We are also in something close to verifiable contact with our own dead, or something that appears to be that.

It would seem that, as we get closer to the extinction crisis, the hidden world, which in the past has always been vague and hard to classify, is coming into a new sort of focus —is, in fact, turning out to be real.

Except…exactly what does that word "real" actually mean?

Come with me, to the rabbit hole where we now must go.

I know that my stories are generally impossible to believe. But I also know that they happened.

Or do I?

The world of the visitors is so strange, so completely out-of-bounds, that one would think that surely it cannot be factual. If it is, then what we now consider real is the actual fantasy.

Or could they *both* be real and fantasy at the same time?

Let's explore.

A growing body of theory and now also evidence suggests that there may be no final truth, not in this universe. Perhaps others have different laws, but in this one, belief does not work—and not just on the level of the extremely small—the quantum level, where nothing is fixed and definite until it is somehow noticed. Belief doesn't work on our level, either. The reason is that we don't have an absolute ground of truth that can be relied on completely. As we are now, we simply decide that one thing or another is true.

As I reported in *Transformation*, I once saw a city which was described to me as "a place where the truth is known." At the time, I thought, "Oh, my, there must be such secrets locked away there." Now I suspect that the message was more subtle. It could well be that its residents think using an input strategy rather than an output strategy.

Because we use an output strategy, we are definitely going to know that the structure in the pantry is a box of cereal and so be able to eat breakfast. But we are not going to know what that box actually *is*—which, for us, isn't going to be a problem. Normally. But we have, with our exquisite minds, evolved formulae and constructed machines that enable us to see below the limit where output strategy applies. We can use detectors to see, with our output-wired brains, the hidden workings of the interior world. In other words, we can see the truth.

And what we see is that it's ambiguous!

The city where the truth is known was flooded with bright light—presumably symbolizing its clarity—but it went on forever. I never came to a border. The reason is, of course, that something that is ambiguous has no border.

We know that quantum indeterminacy is real because we can see it. But, due to the way our brains are constructed and therefore the way our minds work, it makes no sense. A wave can't resolve into a particle just because we're looking at it. But it does. The double-slit experiment that proves that this is true works every darned time.

It also drives me nuts, and I love it.

There are, however, some things about the double-slit experiment that need to be unpacked a bit before we assume that the popular notion that an observer is required for a wave to resolve into a particle is correct. There is a quantum theory called decoherence which suggests that the wave function never collapses, but rather the appearance of coherence comes from its surroundings. Stretching the theory a bit, this might actually mean that our brains are what cause the world to appear coherent to us. In other words, our output strategy is, for us, reality.

But wait a minute, a brain strategy isn't any such thing. It's just a method of making things look a certain way. It's not even the only output strategy. Bees, for example, who see into the ultraviolet, have a very different one.

While this doesn't solve the mystery, it does suggest that, if the underlying wave function never actually collapses, then its inherent ambiguity is still there even though what we see all around us seems entirely coherent.

And indeed, the Schrodinger's cat hypothesis hints that what we really live in is an ambiguous reality insofar as it predicts that two contradictory states of something can both be simultaneously real. Schrodinger's cat can indeed be alive and dead at the same time. But surely these are physics issues, not life problems for us to worry about. Quanta are very tiny, so they don't apply in our big world of absolute realities and reliable output strategies. Schrodinger's cat isn't any sort of feline at all, she's a subatomic particle. So to quote a great genius of the 20th century, Prof. Alfred E. Neumann, "What, me worry?"

Oh, but I do!

This is because the indeterminacy of the wave function does not appear to be confined to quanta. We're living in it, which is why I call this world of ours a labyrinth of mirrors.

So, it's all ambiguous. Problem solved.

But no, that's not true, either. This is because this whole ambiguous, deceptive and confusing garden of forked paths functions as it does and appears as it is because of...constants.

Oh well…

When I was in the throes of being driven even crazier than my baseline by writing this book, I turned to my implant for help. And people wonder why I don't get it removed. (But why should I, it's not there, after all.) Anne was right, and thank all the holies that her argument prevailed, because, despite the fact that it doesn't exist, it is my trusty guide in the labyrinth of mirrors.

I asked it to lead me to something that I needed to know for this book but knew absolutely nothing about.

Enter physicist Wolfgang Pauli, psychiatrist Carl Jung and the strangest thing in physics, which is something called the fine-structure constant. It is set by nature at 1/137. It's an absolute. If any other fraction was there, everything would be different and most of the world wouldn't work. All well and good, but why that particular fraction? Why not 1/136 or, say, 1/7000?

Here's the fun of it: Nobody knows.

It's just…there.

The other constants can be seen to be inevitable outcomes of the processes that they define. Not the fine-structure constant, though. It's dimensionless. This means that, no matter what system of units you use to derive it, it's always going to be 1/137.

Physically, it's the splitting of the spectral lines of hydrogen atoms.

Which would seem to be no big deal, except for one tiny detail, which is that if that distance was anything else, the world would be entirely different and under most scenarios wouldn't work at all.

And yet, it's just there. There's no underlying principle that compels it to be what it is. God might as well have said, "Hey, this one looks nice. I'll use it."

This drove Pauli crazy, as it has many a physicist since.

I'm not even a member of that particular lunatic fringe (mine is nowhere near as weird), and it drives me crazy, too.

Like Pauli, I love things that do that. Anne always said during her lifetime that tackling unanswerable questions makes the mind grow. I think the mystery of the visitors is just this sort of question, and maybe they are keeping themselves so ambiguous in order to induce an increase in human intellect.

Even though they're answerable enough in the output world, go inside and all is ambiguity. But that's ok. We're in the output world. An apple is an apple here, no mistaking that.

Except...well, I've been warning you.

The great Pauli, still being driven crazy by the why of the fine-structure constant, died on December 15, 1958...in room 137 of the hospital where he had been admitted with an advanced case of pancreatic cancer. Not only did he die with his question, he did it while entangled in one of the mysterious synchronicities that his friend Carl Jung writes so much about. (Unless it was a setup and he had himself moved to that room on purpose. Could be, he was a tricky guy...)

My nonexistent implant uses synchronicities, as I have said. In fact, it used one to draw me to this very mystery. But don't forget: it's not there. There's no there there anymore. Except, of course, Schrodinger's cat is still alive, too...

How ironic that the one constant on which deliverance from ambiguity most depends seems to have come out of nowhere.

In fact, just like the input strategy, the truth is that our precious, stable, coherent output strategy also produces outcomes that are real and illusory at the same time. The input strategy starts there, because it must start with indeterminacy.

But our strategy starts out very definite. As I said, an apple is an apple.

And now let me introduce another prankster. In fact, a philosopher. (No, please, don't go into a coma. This is one terrific philosopher. In just a few words, he's going to change everything you know about everything and you're going to have loads of fun into the bargain. Unless, of course, you go insane…)

In 1963, Edmund Gettier published a short paper called "Is Justified True Belief Knowledge?" In so doing, he gave birth to what is now known as the Gettier Paradox, which offers a challenge to the idea that something that is demonstrably true is also final and absolute knowledge. He showed that what is known among philosophers as a justified true belief cannot exist. In other words, he met the demand of an earlier philosopher, Friedrich Nietzsche, that there must evolve a "philosophy of the dangerous perhaps." What Gettier demonstrated is that, even if you are certain from observation that something is true, it cannot, in a final sense, be knowledge.

Enter the cow problem. A farmer has a Holstein who is showing signs that she might be ill. He decides to isolate her so he can watch over her, so he takes her into a field that he can see from his barn while he works. The field has only two features in it, a tree and a small hollow. Only if the cow wanders into the hollow will he be unable to see her. This is so unlikely that he's not concerned about it.

He goes about his work, from time to time glancing into the field and observing the cow, who appears to be all right. While he's oiling his tractor, she wanders into the hollow. At the same time, a random puff of wind blows a piece of black and white paper into the field, which gets caught under the tree.

So this is now the situation: The cow is in the hollow

invisible to the farmer. She is all right. There is a piece of black and white paper caught under the tree.

He looks into the field and sees the paper. From this distance, he can't see detail, so he assumes that it's his black and white Holstein and concludes that she's all right.

He is now both right and wrong at the same time. He thinks he saw his cow, but he didn't. So he's wrong. But she is indeed fine. So he's right.

As we observe the world around us, and even though our output strategy works time and time again, we can never know, in an absolute and final sense, if we are right about anything we observe.

And yet, we're surrounded by a world that seems to be completely true. But this truth cannot, in fact, be knowledge. Add to this the problem that we don't actually see it.

It would seem that reality is not fixed at all, but more like an ocean that never stops moving. But it is also a sea of wonder. We even have a name for it. Many names, in fact. But the most important one for us is the one closest to home. We call this one "the human mind." It is also the universe, for everything we see is of necessity inside our heads. For us, there is no outside. The whole world and all experience is and must always be inside us, in the singing, sputtering neurons that cradle our minds. We see only what our detector —our brain and its various input devices such as eyes, nose, skin, ears and so forth—delivers to us. This is never what is really there, but rather what our brain is able to see. We never observe the shimmering ultraviolet world that the bee does.

The brain is our detector. As we are curious sorts, we have for a long time been extending its reach with prosthetic devices. And what an extraordinary story they tell us. The optical prosthetics we know as telescopes reveal wonders in the sky...and the mysteries that they reveal, again and again, confirm that the one thing we must believe in without question is question itself.

Ambiguity rules, yes, but is that true everywhere? Maybe there's some little corner of the universe where constants produce absolutes like it would seem that they should, not all the darned questions we have to cope with.

We recently used one of our prosthetics to find out. It's really two of them being operated together, the William Herschel Telescope and the Telescopio Nazionale Galileo, both located on the island of La Palma in the Canary Islands. Using the two scopes, a team of scientists led by quantum physicist Anton Zeilinger from the Austrian Academy of Sciences looked at quasars located at opposite ends of the universe. They collected photons from these quasars then entangled them by bringing them together in yet a third machine, a mobile laboratory located at the nearby Nordic Optical Telescope. (There are six telescopes on La Palma.) They were then sent to receiving stations near the other two scopes, and their entangled state was determined.

It was found that light gathered from the opposite ends of the universe can be entangled just like light from two different lamps in a lab, meaning that the laws of quantum physics are consistent in one important way: they are true from one end of the universe to the other.

Telescopes were used to gather light—classical physics at work, which we understand perfectly well...sort of. But the light was then used to induce quantum entanglement—which makes no sense whatsoever.

The problem is that both physics work, albeit at different levels. But the comfortable assumption that things make certain sense at the classical level is probably wrong, as the Gettier paradox so eloquently illustrates.

I love it! In fact, just for fun, I paused for a moment and went into the john where there's a white wall, at this time of the morning brightly lit by the sun. I said to my implant's ghost, "Whatever is right is right, yes? Or no? Or yes and no?" Sure enough, after a few seconds I saw words racing past in the slit that opens in my eye. After concentrating for a few moments, I managed to follow a phrase, "city may emulate Apophis," and then I saw the word "concentration" before the stream sped up once again.

When Apophis flew past, I laughed out loud. This is because, in the Egyptian pantheon, he is Chaos, the opposite of Truth! So what am I to think? Has my implant perhaps just told me that the city where the truth is known is abandoned, perhaps even in ruins? If so, that could be a problem. Because, as you may have realized, it's not what I thought when I was writing *Transformation*, a marvelous city on some distant planet. It's here. We live in it. It's there, too, of course. The city where the truth is known is the universe itself.

We have explored many of its dangers over the course of this book, but its greatest danger is ignoring the fact that the truth has no resting place.

If we come into communion with the visitors, we are going to start sharing the rigors of their input strategy just as they are the delights of our output strategy, at the same time, living in the deeper reality that the truth that is known is that there is no final truth, not anywhere in this vast cradle that we all call home. But there is a certain potential for balance, and that is where the most intimate and richest opportunity for communion is to be found.

What about this balance, though? What would balance between us and the visitors actually be? How would it work and feel in our lives?

Before Anne and I were married, she crocheted a sampler which hangs on our bedroom wall to this day: "Two are better than one, and a cord of three strands is not quickly snapped."

These lines from Ecclesiastes, sewn to the sampler when we were just getting started together, turn out to be the foundation not only of married life, but also of relationship with the visitors, and the final theme of this book. We are one strand, they are the other, and the communion that is seeking to occur between us is the third.

It's worth seeing this in terms of Gurdjieff's three forces. We are the passive force, trying to understand how to open ourselves to them. They are the active force, trying to find a way to enter us. The third force is the whole mystery that lies between us, the fear, the longing, the curiosity, the wonder, the danger, the joy, all of it. In reconciling these many different states and finding the balance that underlies them all, we will find communion. In the end, simply this: the love that underlies life, which Anne calls "the yearning," is as much a mystery as the fine-structure constant, something on which our hearts depend, and maybe, like the fine-structure constant, also all that is.

Right now, both sides are making a choice. They are deciding to keep us or let us go. We are deciding whether or not we can bear them inside us. If my own experience with them has any validity at all, it is that my journey from the terror of a wild animal to the excitement of discovery that is my life now is well worth the risks and the struggle.

Standing in the way of our drawing closer, there is the matter of unsureness. Are they parasitic, here to take without giving, or symbiotic, here to share? If they are parasites, then they are the greatest danger we have ever known. If they are here to share, though, then they are literally the pearl of great price, worth giving everything we have to acquire.

There is enough dark material to justify hiding from them or trying to fight them. There is enough outlandish testimony to justify rejecting it all and calling the whole thing ridiculous. There is enough ambiguity about the state of the earth to decide that the climate emergency isn't going to lead to anything close to an extinction event and, therefore, that there's no urgency here.

But the greater preponderance of evidence is that they are real and here and that our situation is indeed urgent and that we can find a new way and a new world together.

If they—and we—both draw a little closer, more opportunities for communication will follow, and there will be a more general offering of communion. I know the visitors too well to predict exactly how—or if—it will all unfold. But if I have done my job correctly, they will move the bar a little bit.

I wonder what it will mean. Are we standing in the jaws of a trap that is waiting to be sprung or on the edge of glory? Or is the truth that such a place in history must always be both?

I hope that this chapter has made it clear: Just as Schrodinger's cat is both dead and alive at the same time, this entire experience is both a trap and the key to unlocking it.

Accept the question, accept the truth. Live the question, and you are living in the new world.

## 13

# IT IS TIME

Throughout this book, I have made what I suppose is rather annoying reference to "the visitors" with very little effort made to differentiate between forms. This has been quite purposeful. The reason is that I don't know what the forms mean. I have seen the *kobolds*, the goblin-like grays, the tall blonds, some human-like creatures who could read minds and seemed very disturbed, another form that I would describe as being like brilliant little primates. These are the ones who swarmed into our apartment in 2007, whom I mistook for dogs, and who seem to have been with me since. I have also seen a dead man come into physical form, and I have seen normal-looking human beings who appeared to be somehow part of this whole enigmatic web.

The ones that I have seen while in a normal psychological state and while they were entirely physical in appearance are a child of the tall blonds, the various humans, the *kobolds* Anne and I saw in Manhattan, the people who entered my house and placed the implant in my ear, the short disturbed person and his two ghoulish companions and the Master of the Key.

I have observed a complex and nuanced range of approaches from all of them. What's more, there is no such thing as a reliable narrator in this experience. As we are now —and as they are—that cannot happen. Perhaps somebody in the future, with more facts to draw on, can provide a more accurate assessment than this. Right now, all I can do is tell my stories without any more attribution than the memories themselves provide.

Anne always warned me, "Don't connect the dots" until I had provable reason to do so. I don't, so I won't.

All of that said, I have amassed, over the years, a considerable amount of information about what our visitors think of us and our world, and much of it is quite different from how we think of ourselves.

First, they do not see the fantastic human bloodletting, which started in 1914 and has now taken something close to half a billion lives, in terms of political and ideological clashes.

Insofar as I understand what I have learned from them, all of this killing has more to do with population pressure than politics. Like many other animals, we can become violent when we feel crowded. As with many other species, in our case, we start killing each other. Hitler wants more living space and creates an elaborate ideological justification for killing others in order to obtain it. Stalin's paranoia causes him to see millions of people as supernumerary, and he starves whole populations to death. Mao does the same. Meanwhile, the United States, not under the population pressure of Europe or Asia, remains peaceful.

So far, to the visitors, this all seems quite natural. But then the atomic bomb appears, and they realize that we could go too far. We could commit species suicide and die out entirely. They therefore begin a process that is intended to lead to contact, at which time, they hope to intervene techno-logically in order to, in effect, expand the effective size of the planet with innovation, thus making each of us feel a greater sense of space around us and stopping the instinctive self-destructive impulse that is threatening to destroy us.

But because of the complex gulf between us, which I have endeavored to make a little clearer in this book, they do not emerge immediately. Instead, they begin a process of social engineering based on contacts with some governments and the formation of human social organizations with specialized connection to them that are intended to eventually bridge the gap.

Then the 21$^{st}$ century arrives, and the American popula-tion, having exploded from 78 million in 1900 to 230 million in 2000, begins to feel the same sort of instinct toward self-destruction that caused the bloodlettings in Europe. But we are different and so are the times. We have the ability to destroy life on Earth with the push of a button. We also hold a world leadership position when it comes to climate change. Between 1970 and the present, American society becomes more and more involved with the death wish until, in November of 2018, a devastating climate report called the Fourth National Climate Assessment is released by govern-ment scientists, and the US president responds, "I don't believe it."

In my own life, the response from the visitors is immedi-ate. They increase their engagement with me, becoming more and more strident that I provide what they regard as some-thing I volunteered to do and led them to believe that I could do, which is to write a book that will describe enough about the experience of communicating with them to enable them to usefully widen their contact with us.

I end up in the unpleasant position of being an author

without any hope of finding a publisher, who is under extraordinary pressure to get a book published in a matter of months that he doubts it is even possible to write.

So here we are.

But that's not quite the end, for there is another level of this that requires exploration before this is finished. It has to do with what we are and what our place is in the much larger world in which we find ourselves.

We may not be what we seem. In fact, the form we are in just now might not even be the only human form. One of the most familiar alien types may actually be…us.

You think not? Read on.

A psychologist I knew told me a story of driving along the Grand Central Parkway in Queens, just passing LaGuardia Airport. To his horror, he saw a jet descending onto the highway. At first he was terrified, but as it passed over at low altitude, he realized that it was not a plane at all. As it sailed over his head, he thought that it was some sort of gigantic cardboard cutout. Of course, he was extremely puzzled but also in traffic and so couldn't keep watching it. As he looked away from it, he noticed a signboard on the roadside flashing letters that appeared to him like hieroglyphics. Then he realized that there were cars parked along the shoulder and that a number of people were standing in a circle just off in the shadows. Naturally, he wanted to know more, so he worked his way over, parked, got out and approached them. As he did, a small, dark figure came up to him and snarled, "Get out of here."

He obeyed. But what was happening? Was he seeing aliens gathering a certain group of people on the roadside for some inscrutable reason? And what about those hieroglyphics. We can't read them.

Or can we?

When I heard that story, I realized that it and two others

had combined to give me a clue not about us but about the *kobolds*, and it is a surprising one indeed. I remembered both the time Anne and I had seen similar creatures in a little storefront in Manhattan in the process of kidnapping the frantic man in the suit and a remarkable story from Lorie. She had been lying in bed one night about 11 in 1954. Her husband was working, and she was home alone. She was pregnant. She noticed movement and glanced up. To her horror, a line of short, dark blue men with wide faces were standing beside the bed. As she recoiled in horror, the one in the lead said, "Do not be afraid. We're not here for you. We're interested in the girl child you're carrying." Of course, this absolutely terrified her. It then laid its hand on hers and asked, "Why do you fear us?" She blurted out, "Because you're so ugly!" When I heard his response, many puzzle pieces came a little closer together for me. He said, "One day, my dear, you will look just like us."

About ten years after hearing this story, I was in a surgery waiting room with dear friends. The father of this family was undergoing emergency heart surgery. It was unlikely that he would survive. The next thing I knew, I saw him come walking out of the operating theater. At that point in my life, I had seen many dead people appear like that, and I whispered to Anne that he had just died. A moment later, the surgeon appeared and gave his wife the news.

I then saw, in my mind's eye, something quite extraordinary. He was being carried by two *kobolds*. They were not here, but in a crowded forecourt in what appeared to be India. They made a quick sort of turning motion, and he was inside the body of a very new baby that a young man was holding in his arms. The baby erupted into screaming, and I knew at once what had happened.

He had not been a very good man in this life but, at the same time, had raised some fine children who were a real credit to him. If he had had to face his life, he would have fallen at once and all the good that was there buried in him would have been lost. He deserved another chance, which is why he had been raced into this new body before he had a chance to really see himself.

When the children would see the *kobolds* at our cabin, they would say that they called themselves doctors. They would shine lights on them and tell them that they were examining their souls. Lorie Barnes heard them call themselves "soultechs." Soul technologists, I would think.

I know that all of this sounds fictional. Of course it does. I don't think it is, though. I think that I am describing things that we will see and understand more clearly as we draw closer to the visitors and thus also to our own real place in the world.

What these stories tell me is that the *kobolds* are in some way part of us, perhaps another form of the human species that manages the movement of our souls and watches over them.

Many species have more than one form. It's not unusual in nature for a larval form and a mature form to be completely different, as the tadpole and the frog and the caterpillar and the butterfly show.

As science advances in its ability to see reality and philosophy in its ability to interpret its vision, it is coming to seem that the world is far larger and more complex than we ever imagined, and mysterious beings like our visitors, which we dismissed as imaginary, are turning out to be in some ways quite literally more real than we are. Communion with them seems not so much to be something to be attained but something that has always been part of us, and which is now raising its ancient voice again, calling to all of us to look up to the stars and inward to the infinity of our minds, and find our own greatness, and save ourselves.

When the visitors draw close to you, barriers like vast the distances between the stars come to seem almost trivial. As they slip in secret through our skies and our thoughts, they become less like mirages, and more like an eerie truth that we have thus far been afraid to face.

As we are coming to realize that they are real and here, we also find ourselves asking the same question that Col. Philip Corso asked so long ago when he found himself facing them in the darkness of a cave: "What's in it for us?"

The bottom line here is this: whoever they are, and whoever *we* are, they want to join us and live in conscious contact with us. They need us, but more, we need them—their wisdom and their devastatingly accurate insight into the fragile truth of the world.

Dare we open our door to them?

On the night of Wednesday, June 12, 2019, I sat to do my usual 11 PM exercise. Almost immediately, I felt a sense of pressure all around me, as if the air had become more dense. It was almost as if I had been wrapped in an invisible blanket. The next thing I knew, I heard a young male voice say excitedly, "We're in!" The next instant, I saw hanging before me a complex schematic. I found myself pushing it away with my mind, which caused it to vibrate. I did it again, but then the thing disappeared, and I knew that somebody had just forced their way into me. (I cannot say what was on the schematic. I have no idea what it was, only that it was some sort of chart that looked like circuitry. I was seeing, I would think, the input level of whatever was there.)

After a moment, I calmed down. I completed the meditation and went to bed. During the 3 AM meditation, I felt my ear get hot. The implant had turned on. A moment later, I found myself wanting to pray for protection. I had been afraid when I was entered. I didn't want it to happen again. On Thursday and Friday nights, I once again surrounded myself in my imagination with protective light, saying that nobody could come through it unless they had something to give me that would strengthen me in return for their participation in my life. I also prayed, invoking Jesus, whose journey I have come to understand in a deep way and whom I have reason to believe can be addressed and will respond.

Once again, even though my door was open, nobody came to me.

Then came June 16, 2019, Father's Day. The time was approximately 4 in the morning. I was in a hotel for the night. I'd done the 11 PM meditation as usual and had just sat down to do the early morning one. As I settled myself into a chair, I noticed a vertical oblong shape, absolutely black, hanging in midair. It was across the room, maybe ten feet from where I was sitting. Wondering if it was a shadow, I moved my head from side to side. It remained stationary. I then felt a sense of presence, exactly as if it was a person standing in front of me.

But this was not a person as we understand that term. It most resembled a vesica piscis, the intersection of two over-lapping disks. It appears in Euclid's first proposition, where it is used in forming an equilateral triangle. It is also both a Christian and a Masonic symbol. In Medieval art, it was used to enclose figures of saints and of Christ. The lid of the Chalice Well in Glastonbury, England, contains a vesica piscis. In ancient times, the area enclosed by the vesica was more than just empty space. It was a sacred entity. This is why sacred figures were placed within it. The reason that it was considered sacred was that, before the development of sophisticated mathematics, geometry was used in the plan-ning of structures. It was the foundation of human endeavor, and the vesica was the fundamental geometric form where all measurement began. Man built using geometry, and the vesica contained the basis of measurement. This is also why it is an important form in Masonry.

In my hotel room that night, then, was a living representation of this fundamental and deeply sacred form. It was linked to me through the Chalice Well, where I have experienced some of the deepest and most joyous times of meditation in my life.

I wish that I'd had the presence of mind to understand all this in the moment, but I was excited and pretty amazed, so I really spent my time with the form watching it as it moved toward me. The next thing I knew, it had dropped to the floor and was lying across my feet. I felt something light but solid there. It seemed like a living thing, although I don't think that it was a biological entity as we understand such things. It was more of a living symbol, in some way guided by intelligence. Or maybe it was actually a body that I was seeing though an input strategy and therefore observing what it is before it focuses into a physical form. Had I seen it through an output strategy, maybe it would have appeared as a creature.

I looked down, and the vertical oval now covered my feet with blackness. I thought at once, "If I don't move, it will enter me."

Another moment had arrived like the nine knocks, the confrontation in the woods in 1987 and the time the entity got between my legs in 2017. All three of those times had caused a fear response and a refusal of the call.

This time there was a very different feeling. It felt as if the entity lying across my feet was beseeching me for entry into my body. What would that mean, though? I thought perhaps it would be too much to give up to let it into me so frankly and directly. It is one thing to commune during meditation, another to allow the actual, physical penetration of another consciousness. Much more of a challenge, to say the least. I had the sense that it wanted badly to do this, but also that it was begging that it would not force itself on me.

I moved my feet back slightly, and it at once disappeared.

For a time, I continued with the sensing exercise. The next thing I knew, I was in a sort of waking dream. There was a man pointing a big silver gun at my face. However, it wasn't real. I could see clearly that the barrel was plugged. In fact, it looked like the sort of toy gun I used to play with as a child.

The event then ended. It was once again just me, the shadowy hotel room and the distant hiss of night traffic on the highway outside. I continued the meditation for a short time then returned to bed. I slept the limitless sleep of a child.

To me, this communication, coming just as I was finishing this book, meant that they want deeper communion —in fact, that there is a demand there, although a gentle one.

They are not going to force anything, and there is also an element of play involved. Thus, the use of a child's toy that represents a lethal weapon.

Back at home the next night, I felt regret for the conditions I had placed on the entity. Again, I asked for guidance.

The next night, things returned to normal. Before the early morning meditation, there sometimes appears hanging in a line on the window beside my bed a series of hieroglyphics. They are not Egyptian and are generally only there for a moment. I do not understand them, and so far I cannot say that I have seen the same sequence twice. This happens only a few times in a year, and it happened that night. What it signifies, I don't know, but as soon as I saw it, I went to do the meditation, which became so deep that it was as if I was sitting in another world and in this one at the same time.

I let my body be open to the need I sense is there.

If I had continued to refuse, would the plea have become more forceful? Would the gun have become some sort of genuine threat? It's a question I cannot answer. I can only forge on, trusting myself and the visitors as best I can, never forgetting that communion is not about doing something new, but accepting something that has always been part of human life by bringing it into conscious awareness. When we do this, there is an exponential leap in richness for both sides. I know this because, on October 16, 2019, I finally allowed entry while I was in full consciousness. Not only that, I managed, quite by accident, to record the moment in audio.

As I've said, despite years of trying, I've had very little luck with video. On the afternoon of the 15th, I happened to see that there was such a thing as a "sleep recorder" app that records sounds made while you sleep. I thought that maybe I could experiment with this. When the evening of the 15th arrived, I sensed that the visitors might make a close appearance—not necessarily physical, but very close. After an extremely deep 11 PM sensing meditation that lasted the better part of an hour, I went to sleep. After the 3 AM meditation, as I was falling back to sleep, I felt weight come down onto my legs. Unlike what happened with the vesica piscis, I felt within me a calm sense of welcoming. I was open. Then I fell asleep.

In the morning, I listened to what the sleep recorder had picked up. To my astonishment, amid the predictable grunts and snores, there were some words. Just after 4, I say in a sleep-dense voice, "What is that?" Then, a second or so later, still coming up from sleep, "Oh, Mature." Another few seconds pass and I say in an entirely different voice, rich with pleasure and anticipation, full of sensual joy, "Teach me, Mature." A short time later, there is a little sigh that does not sound like me. It sounds feminine. A sound expert has analyzed it, although not too deeply, and also feels that it is feminine. But there was no woman there.

Or was there? After about fifteen minutes of silence, I am heard grieving for Anne.

Let me explain what I understand of what happened. First, the use of the word "Mature" as a name. I think that it is used in place of "Master" because that word is too ego-charged.

My sense is that the visitors prefer a word like Mature, which indicates somebody who has no need to drop into the physical world again. In other words, what we would call an ascended master.

Except for the sigh, the entity never makes a physical sound. In my experience, there have been almost no physical words. "Have joy" and perhaps a few others. What I remember is a sense of intense intellectual contact causing pleasure that was almost sexual in nature. If you can imagine an exploration of ideas so beautiful and intense that it was like a form of spiritual sex, that might be in the direction of what I felt.

I think that this whole book might have been examined in those few moments. I think that, unlike *Communion*, which they more-or-less laughed at, it was accepted as a useful effort. (You will find the story of their reaction to *Communion* in *Transformation* and *Super Natural*. Suffice to say that they let an editor at William Morrow & Company know that I'd gotten a lot wrong, and found the book nothing more than amusing…but, of course, at the same time they honored my effort by appearing in front of the editor right in the middle of a bookstore, no less.)

I must not get this wrong, not this time. We urgently need to rise to a new level of coherence, which we cannot do except in deepest communion. And by "we" I mean all levels of entity, physical and other, human and other.

On the morning of October 27th, 2019, I had an encounter that felt physical and also ambiguous—in other words, an encounter with what life in communion will be like.

You will recall the short man who lurked in the woods in upstate New York and then outside our condo in Texas. I saw him again, but this time in a very much better state. He looked clean and calm and happy. His eyes were shining. He was not smoking, and opened his mouth and showed me the pink tongue of a child. Then I saw a woman with a lovely, sweet face. I couldn't tell if they were physical or not. I did not seem to be dreaming, but they were also not plainly standing beside the bed. They were in a place that will, I suspect, become more familiar as we go deeper into communion, which is the quintessential "dangerous perhaps," the uneasy edge between our ordinary world and the greater mystery in which it is embedded. The woman seemed at once a stranger and yet also very familiar to me, and seeing her inspired a poignant sense of memory. I asked her, "How long have you known me?" She replied, "Since you were born."

And then it ended. I slipped into sleep and they into the ocean that tosses beneath the lives of us all. He is transformed, that poor boy who followed me in the mysteries of the night, a very real person in possession of tools of mind that enabled him to enter into places in me where he did not belong. Now, as I have found peace, his pain has also released its grip on him and he has become a tender child.

It's impossible, of course. He would be thirty now, or more. Nobody like that could live very long. Somewhere in some shadowy place, I am sure his bones lie, maybe buried, maybe not. And her, what of her? All in white, she stood, a girl like a drop of sunlight, a young mother I thought, and I felt her as a mother. But she cannot be mine.

In those few minutes, I was under the water of life, in a place that communion must inevitably take us, where our demons and angels dance together, sweetly singing and beckoning us to walk the cliffs of knowledge, ever closer to the edge.

There it ended, and I think that this lovely, gentle experience perfectly illustrates what I have been suggesting throughout this book, which is that, if we are to proceed beyond contact and into communion, we are going to have to embrace the fundamental ambiguity of consciousness itself, and accept the mystery that is all around us as the "immense great benevolence" that the attendee who saw me briefly disappear at the conference perceived during her experience.

It is morning now. There is bright sun. I can smell smoke from distant fires. Later, I will begin another night's journey, open now to embrace and no longer afraid. Night before last, as happens often now, a presence settled down on my body. This time, though, it had more weight and substance than ever before. I opened my eyes but could see nothing. I could only feel it there. The weight was solid and very real. I was open and ready. But it gently lifted away.

In just these few months since the encounter with the vesica piscis, I feel that I have finally gone beyond fear. I see how the boundaries with which we define ourselves are its real source, for they don't actually exist and we know it, and facing our borderless, wandering reality feels very close to ceasing to exist. Communion, in this sense, feels like death, which is why Jeff Kripal was so devastated by even a brief second of it on that night at Esalen. But we have no boundaries. There is only one of us—alien, human, living, dead, whatever. What we are is a wave-front of consciousness, speeding into the unknown.

When you embrace this ambiguity as yourself, you discover the abiding peace the Hindus call *shanti*. It makes no sense. It reconciles nothing. And yet everything within you, all the fears, the angers, the hatreds, the lusts, the disappointments, the ambitions, all that lies within the scope of your life, comes to rest, and you know that you have found your heart.

It is what Anne meant when she said, "Enlightenment is what happens when there is nothing left of us but love." When all that you have been fighting against and for is stripped away, your nakedness that has so frightened you for so long, is soothed by the gauze of angels.

It's not easy, though. The first step out of oneself and into communion is a very hard one to take. Open, innocent surrender to the enormous presence that underlies reality is never going to be easy, and it is never going to be certain. But it is also a priceless resource, offering a path into greater knowledge, a new science that is more true because it includes more of what is real, philosophical understanding that feeds the mind with the stuff of truth, and limitless expansion of the scope of mankind.

This series of interactions I have just described is a perfect example of how communication with the visitors unfolds. I asked a question and was answered with a series of demonstrations. The first manifestation forced its way into me. The second beseeched, then suggested that a threat was possible if I continued to refuse it entry. I felt an acute sense of failure and sought to open myself. When it came back this time, I did not resist.

And now here I am face to face with what seems to me to be something of great beauty, at once real and unreal, a challenge, as it were, from Stevenson's land of counterpane, the sweet winds of Asphodel singing to my heart.

Still, practical, crucial questions remain. Will the visitors come closer? Can they ever be physically real enough to matter as more than hypothesis? In other words, will the conjuring that is this book work? If so, will they turn out to be impossible for us to bear, or will they become in general life something like the engine of joy that is beginning to seek toward me?

If these questions are to be addressed usefully, the visitors cannot just come to me, sly in the whispering night, not anymore. The visitors must open the doors of their school wide, to us all. We have a planet to lose and our lives along with it, or we have a journey to take.

Which shall it be? The decision belongs to all of us and

each of us, and to them. Shall we join in what is essentially a new world and a new way of life, or do they disappear into the dark and we into the storm?

We must decide, and now, and so must they. It is time.


The End

# BIBLIOGRAPHY

Berliner, Don, *UFO Briefing Document: The Best Available Evidence,* Doubleday, New York, 2000

Conroy, Ed, *Report on Communion,* Wm. Morrow, New York, 1989

Eliade, Mircea, *Shamanism: Archaic Techniques of Ecstasy,* Princeton University Press, Princeton, 2004

Hopkins, Budd, *Intruders,* Random House, New York, 1987

Hernandez, Rey, *Beyond UFOs: The Science of Consciousness & Contact with Non-Human Intelligence,* Create Space, New York, 2018

Hynek, Philip, Imbrogno, Philip, Pratt, Robert, *Night Siege: The Hudson Valley UFO Sightings,* Llewellyn Publications, Woodbury, Mass, 1998

Jung, Carl, *The Red Book,* W.W. Norton, New York, 2009

Kripal, Jeffrey J., *The Flip,* Belleview Literary Press, New York, 2019

Krohn, Elizabeth and Kripal, Jeffrey, *Changed in a Flash,* North Atlantic Books, Berkeley, 2018

Lammer, Helmut and Marian, *MILABS: Military Mind Control and Alien Abduction,* Illuminet Press, Atlanta, 2000

Marden, Kathleen, *Extraterrestrial Contact ,* Red Wheel, New York, 2019

Mehust, Bertrand, *Jesus Thaumaturge,* InterEditions, Paris, 2015

Morrow, Susan Brind, *The Dawning Moon of the Mind,* Farrar, Straus and Giroux, New York, 2017

Mullis, Kary, *Dancing Naked in the Mind Field,* Pantheon, New York, 1998

Narby, Jeremy, *The Cosmic Serpent,* Tarcher/Putnam, New York, 1999

Ouspensky, P.D., *In Search of the Miraculous,* Harcourt Brace, New York, 1949

Ring, Kenneth, *The Omega Project*, Willam Morrow & Co., New York, 1992

Strieber, Whitley and Anne, *The Afterlife Revolution,* Walker & Collier, 2017

Strieber, Whitley and Anne, *The Communion Letters,* Walker & Collier, 2003

Strieber, Whitley, *Breakthrough,* Harper-Collins, New York, 1995

Strieber, Whitley, *Communion,* Beechtree Books, New York, 1987

Strieber, Whitley, *Confirmation,* St. Martin's Press, New York, 1998

Strieber, Whitley, *The Key,* Tarcher/Perigree, New York, 2011

Strieber, Whitley, *Solving the Communion Enigma,* Tarcher/Penguin, New York, 2011

Strieber, Whitley and Kripal, Jeffrey J., *Super Natural,* Tarcher/Penguin, New York, 2016

Strieber, Whitley, *The Secret School*, Harper/Collins, New York, 1997

Strieber, Whitley, *Transformation,* Beechtree Books, New York, 1988

Vallee, Jacques, *Passport to Magonia,* Henry Regnery, Chicago, 1969

Webb, Don, *Uncle Ovid's Exercise Book,* Fiction Collective 2, Austin, 1988

Wilhelm, Richard, trans., *The Secret of the Golden Flower,* Harcourt Brace Jovanovich, New York, 1962

# ABOUT THE AUTHOR

Whitley Strieber is the author of over 40 books, including the Communion series: Communion, Transformation and Breakthrough, and now the final book in the series, A New World.

Among his notable nonfiction titles are The Key, Solving the Communion Enigma and Super Natural: A New Vision of the Unexplained.

His most famous novels are The Wolfen, The Hunger, Warday, Superstorm and The Grays. His novel series Alien Hunter was made into the TV series Hunters for the SciFi Channel, The Wolfen and The Hunger were both made into films, as was Superstorm (As The Day After Tomorrow.)

 facebook.com/wstrieber

instagram.com/wstrieber

CPSIA information can be obtained
at www.ICGtesting.com
Printed in the USA
LVHW082347121119
637192LV00019B/847/P